Ric Pimentel
Terry Wall

Cambridge

**Endorsed by
University of Cambridge
International Examinations**

NEW EDITION

checkpoint Maths

3

Workbook

**HODDER
EDUCATION**
AN HACHETTE UK COMPANY

Hachette UK's policy is to use papers that are natural, renewable and recyclable products and made from wood grown in sustainable forests. The logging and manufacturing processes are expected to conform to the environmental regulations of the country of origin.

Orders: please contact Bookpoint Ltd, 130 Milton Park, Abingdon, Oxon OX14 4SB. Telephone: (44) 01235 827720. Fax: (44) 01235 400454. Lines are open 9.00–5.00, Monday to Saturday, with a 24-hour message answering service. Visit our website at www.hoddereducation.com

© Ric Pimentel and Terry Wall 2012

First published in 2012 by
Hodder Education, an Hachette UK Company,
Carmelite House, 50 Victoria Embankment,
London EC4Y 0DZ

Impression number   10
Year   2017

All rights reserved. Apart from any use permitted under UK copyright law, no part of this publication may be reproduced or transmitted in any form or by any means, electronic or mechanical, including photocopying and recording, or held within any information storage and retrieval system, without permission in writing from the publisher or under licence from the Copyright Licensing Agency Limited. Further details of such licences (for reprographic reproduction) may be obtained from the Copyright Licensing Agency Limited, Saffron House, 6–10 Kirby Street, London EC1N 8TS.

Cover photo © Travelscape Images/Alamy
Typeset in Palatino 10.5/12.5 by Pantek Media, Maidstone, Kent
Printed in Great Britain by CPI Group (UK) Ltd, Croydon, CR0 4YY

A catalogue record for this title is available from the British Library

ISBN 978 1444 144 055

Contents

		SECTION 1	**1**
Chapter 1		Integers, powers and roots	1
Chapter 2		Expressions and formulae	7
Chapter 3		Shapes and geometric reasoning	18
Chapter 4		Length, mass and capacity	27
Chapter 5		Planning and collecting data	30
Chapter 6		Calculations and mental strategies 1	32
		SECTION 2	**38**
Chapter 8		Place value, ordering and rounding	38
Chapter 9		Equations and inequalities	44
Chapter 10		Pythagoras' theorem	55
Chapter 11		Compound measures and motion	58
Chapter 12		Processing and presenting data	64
Chapter 13		Calculations and mental strategies 2	71
		SECTION 3	**74**
Chapter 15		Fractions, decimals and percentages	74
Chapter 16		Sequences	81
Chapter 17		Position and movement	85
Chapter 18		Area and volume	91
Chapter 19		Interpreting and discussing results	96
Chapter 20		Calculations and mental strategies 3	98
		SECTION 4	**103**
Chapter 22		Ratio and proportion	103
Chapter 23		Functions and graphs	107
Chapter 24		Bearings and drawings	118
Chapter 25		Measures and the circle	126
Chapter 26		Probability	134
Chapter 27		Calculations and mental strategies 4	138

SECTION 1

1 Integers, powers and roots

Directed numbers

Directed numbers can be positive, negative or zero. They can be added together, subtracted from each other, multiplied and divided. The rules for multiplication and division are the same:

- When both quantities are positive or both are negative, the result is positive.
- When one is positive and one is negative, the result is negative.

Exercise 1.1

Use a number line if necessary to answer questions 1–5.

1 a) (−15.5) + (−9.5) = _____

   b) (−17.1) − (−11.5) = _____

2 a) (−0.3) + (+14.3) + (−21.6) = _____

   b) (+13.7) + (−22.6) + (−20.3) = _____

3 a) (+9.75) − (+6.3) = _____

   b) (+1) − (+1.55) = _____

4 a) (+1.34) − (+2.56) = _____

   b) (+1) − (+15.3) = _____

● CHAPTER 1

5 a) (−7.6) − (+3.17) = _____

   b) (−2.5) − (+2.5) = _____

6 Complete this multiplication grid.
Give your answers correct to two decimal places.

×	−5.2	−3.3	−1.1	0	+2.4	+3.6	+4.7
+3.2							
+2.5							
+1.4							
0							
−1.6							
−2.8							
−3.75							

7 If $xy = +8$, complete this table. (Remember, xy means x multiplied by y.)

x	+8	+4	+2	+1	−1	−2	−4	−8
y								

Powers and roots

Squaring a number is multiplying it by itself. The inverse of squaring a number is finding its **square root**. Every positive number has a positive and a negative square root. To estimate the square root of a number which is not a square number, use the square numbers that it falls between as indicators.
Example 1: $\sqrt{20}$ is between $\sqrt{16} = 4$ and $\sqrt{25} = 5$. So $\sqrt{20}$ is about ±4.5.

Cubing a number is multiplying it by itself twice. The inverse of cubing a number is finding its **cube root**. To estimate the cube root of a number which is not a cube number, use the cube numbers that it falls between as indicators.
Example 2: $\sqrt[3]{250}$ is between $\sqrt[3]{216} = 6$ and $\sqrt[3]{343} = 7$. So $\sqrt[3]{250}$ is about +6.3.

Exercise 1.2

1 Without using a calculator, work out the square roots in parts **(i)** and **(ii)**. Then estimate the square root in part **(iii)**. Give positive and negative roots.

a) (i) $\sqrt{25} =$ _____   (ii) $\sqrt{36} =$ _____

   (iii) $\sqrt{30}$ _____

b) (i) $\sqrt{4} =$ _____   (ii) $\sqrt{9} =$ _____

   (iii) $\sqrt{5}$ _____

c) (i) $\sqrt{0.81} =$ _____   (ii) $\sqrt{0.64} =$ _____

   (iii) $\sqrt{0.7}$ _____

d) (i) $\sqrt{0.64} =$ _____   (ii) $\sqrt{0.49} =$ _____

   (iii) $\sqrt{0.55}$ _____

2 Estimate each of these square roots.

a) $\sqrt{60}$ _____

b) $\sqrt{77}$ _____

c) $\sqrt{0.6}$ _____

d) $\sqrt{0.4}$ _____

3 Without using a calculator, work out the cube roots of the numbers in parts **(i)** and **(ii)**. Then estimate the cube root of the number in part **(iii)**.

a) (i) 64 _____   (ii) 27 _____

   (iii) 42 _____

b) (i) 125 _____   (ii) 216 _____

   (iii) 180 _____

c) (i) 729_____ (ii) 512_____

(iii) 520 _____

4 Without using a calculator, estimate each of these cube roots.

a) $\sqrt[3]{40}$ _____

b) $\sqrt[3]{300}$ _____

c) $\sqrt[3]{20\,000}$ _____

Indices

A short way to write squares, cubes and other **powers** is to use **index notation**.
Example 1: $3 \times 3 \times 3 \times 3 \times 3 = 3^5$

The basic **laws of indices** are:

- $a^m \times a^n = a^{(m+n)}$
- $a^m \div a^n = a^{(m-n)}$
- $(a^m)^n = a^{mn}$

Example 2: Simplify $5^3 \times 5^2 = 5^{(3+2)} = 5^5$
Example 3: Evaluate $(3^2)^3 = 3^{(2 \times 3)} = 3^6 = 729$
Example 4: Evaluate $6^{-4} \times 6^{12} \times 6^{-6} = 6^{(-4+12-6)} = 6^2 = 6 \times 6 = 36$

If the base numbers are not the same, only parts of the expression can be simplified using the index laws.
Example 5: $2 \times 2 \times 2 \times 7 \times 7 = 2^3 \times 7^2$

Two further laws of indices are:

- $a^0 = 1$
- $a^{-m} = \dfrac{1}{a^m}$

Example 6: $12^0 = 1$
Example 7: $5^{-2} = \dfrac{1}{5^2} = \dfrac{1}{25}$
Example 8: $(5^3)^4 \div (5^4)^3 = 5^{12} \div 5^{12} = 5^0 = 1$

Exercise 1.3

1 Simplify the following using indices.

a) $4 \times 4 \times 4 \times 4 =$ _____

b) $3 \times 3 \times 3 \times 3 \times 3 \times 3 \times 3 =$ _____

INTEGERS, POWERS AND ROOTS

2 Write out the following in full.

  a) $12^4 = $ _____

  b) $8^5 = $ _____

3 Simplify the following using indices.

  a) $2^2 \times 2^3 \times 2^4 = $ _____

  b) $15 \times 15^2 \times 15 \times 15^2 = $ _____

  c) $2^2 \times 2^3 \times 2^4 \div 2^2 = $ _____

  d) $7^3 \times 7^2 \times 7 \times 7 \div 7^2 = $ _____

  e) $9^3 \div 9^2 = $ _____

  f) $7^5 \div 7^2 \div 7^2 = $ _____

  g) $5^5 \div 5^3 = $ _____

  h) $3^9 \div 3^9 = $ _____

4 Simplify the following.

  a) $(3^5)^2 = $ _____   **b)** $(2^3)^4 = $ _____

  c) $(7^7)^2 = $ _____   **d)** $(8^4)^4 = $ _____

5 Simplify the following.

  a) $8^3 \times 8^4 = $ _____

  b) $10^5 \div 10^2 = $ _____

  c) $7^4 \times 7^2 \div 7^3 = $ _____

  d) $8^8 \times 8^2 \div 8^4 = $ _____

CHAPTER 1

e) $(2^5)^2 \div 2^3 \times (2^2)^2 \div 2^3 =$ _____

f) $(3^2)^3 \div 3^3 \times (3^2)^2 \div 3 =$ _____

6 Simplify the following. Leave your answers in index form.

a) $4 \times 4 \times 3 \times 5 \times 3 \times 4 \times 5 =$ _____

b) $2 \times 3 \times 2 \times 3 \times 5 \times 5 \times 5 \times 5 =$ _____

c) $3^2 \times 5 \times 5 \times 3 =$ _____

d) $5^2 \times 5^3 \times 3^2 \times 3 \times 5^3 \times 3^2 \times 3 \times 11 =$ _____

7 Using indices, find the value of each of these.

a) $7^4 \times 7 \div 7^5 =$ _____

b) $(2^3)^2 \div 2^3 \times (2^2)^2 \div 2^7 =$ _____

8 Without using a calculator, write each of these as an integer or a fraction.

a) $3^{-3} =$ _____ **b)** $9^{-2} =$ _____

c) $8 \times 4^{-1} =$ _____ **d)** $81 \times 3^{-2} =$ _____

e) $1000 \times 10^{-2} =$ _____ **f)** $128 \times 2^{-3} =$ _____

g) $4 \times 2^{-3} =$ _____ **h)** $9 \times 3^{-4} =$ _____

i) $32 \times 2^{-5} =$ _____ **j)** $2^{-6} \times 2^5 =$ _____

k) $5^{-2} \times 5 =$ _____ **l)** $3^7 \times 3^{-5} =$ _____

Teacher comments

2 Expressions and formulae

Index notation and algebra

The **laws of indices** apply to algebra as well as numbers.

- $a^m \times a^n = a^{(m+n)}$
- $a^m \div a^n = a^{(m-n)}$
- $(a^m)^n = a^{mn}$

Example 1: $m^3 \times m^4 = m^{(3+4)} = m^7$
Example 2: $w^5 \div w^3 = w^{(5-3)} = w^2$
Example 3: $(m^2)^3 = m^{(2\times 3)} = m^6$

If the base numbers are not the same, only parts of the expression can be simplified using the index laws.

Example 4: $g \times g \times g \times h \times h \times g \times g \times h \times h = g^5 \times h^4 = g^5h^4$

Any letter (or number) raised to the power of 0 is equal to 1.

- $a^0 = 1$

Example 5: $w^5 \div w^5 = w^{(5-5)} = w^0 = 1$

Exercise 2.1

1 Simplify the following using indices.

a) $a \times a \times a \times a \times a =$ _____

b) $b \times b \times b \times b \times b \times b \times b \times b \times b =$ _____

c) $c \times c \times c \times c \times c \times c \times c \times c =$ _____

2 Write out the following in full.

a) $p^5 =$ _____

b) $q^7 =$ _____

c) $t^6 =$ _____

● CHAPTER 2

3 Simplify the following using indices.

   a) $a^3 \times a^7 =$ _____

   b) $b^9 \times b^5 =$ _____

   c) $c^2 \times c^4 \times c^6 =$ _____

   d) $e^8 \div e^5 =$ _____

   e) $g^2 \div g =$ _____

4 Simplify the following.

   a) $(m^3)^3 =$ _____

   b) $(p^5)^4 =$ _____

   c) $(m^3)^6 =$ _____

   d) $(p^2)^7 =$ _____

   e) $(x^4)^5 =$ _____

5 Simplify the following.

   a) $a^4 \times a^9 =$ _____

   b) $b^5 \div b^4 =$ _____

   c) $c^4 \times c^2 \div c =$ _____

   d) $(e^4)^2 \div e^5 =$ _____

   e) $(m^2)^3 \div m^5 =$ _____

6 Simplify the following. Do not write multiplication signs in your answers.

   a) $p \times q \times q \times r \times r \times q \times q \times r =$ _____

   b) $t \times t \times t \times t \times u \times u \times t \times t \times t \times u \times u =$ _____

   c) $m \times n \times n \times m \times m \times n \times n \times n =$ _____

   d) $u^2 \times v \times v \times v =$ _____

   e) $s^2 \times s^3 \times t^2 \times t \times s^3 \times t^2 \times t =$ _____

EXPRESSIONS AND FORMULAE

7 Simplify the following using indices.

a) $a^8 \div a^8 =$ _____

b) $b^2 \times b^3 \div b^5 =$ _____

c) $c^4 \times c^2 \div c^5 =$ _____

d) $x^{44} \div x^{44} =$ _____

e) $(p^4)^3 \div (p^3)^4 =$ _____

8 Simplify the following.

a) $n^6 \times n^7 \times n \div n^6 \times n^4 =$ _____

b) $(a^3)^4 \div a^9 =$ _____

c) $(b^4)^3 \div (b^2)^4 =$ _____

d) $(n^2)^3 \times (n^4)^2 \div (n^4)^3 =$ _____

e) $u^2 \times w^3 \times u^3 \times w \times w^3 =$ _____

Expressions

An **expression** represents a value in algebraic form. In the expression $4x - 6$, $4x$ and -6 are **terms** in the expression. Sometimes the expression can be simplified by collecting together the **like terms**.

Factorising is the opposite of expanding brackets. To factorise an expression fully, write the highest common factor (HCF) of all the terms outside the brackets. This might be a number, a letter or a combination of letters and numbers.

Example 1: $20x + 5y = 5(4x + y)$
Example 2: $a^2 + 2ab - 3a = a(a + 2b - 3)$
Example 3: $4pq + 12p^2 = 4p(q + 3p)$

Some expressions do not have a factor that is common to all the terms but can still be factorised by **grouping**.
Example 4: $2ab + 6b + 5a + 15 = 2b(a + 3) + 5(a + 3) = (a + 3)(2b + 5)$

Exercise 2.2

1 Write an expression for the total number of counters in each of these cases, using brackets where possible.

Use r to stand for the number of counters in a full box of red counters, y for a full box of yellow counters, and b and g for full boxes of blue and green counters respectively.

CHAPTER 2

a) 2 boxes of red from which 5 counters have been removed from each box and then the remaining number has been trebled

_____

b) 3 boxes of blue from which 2 counters have been removed from each box and then the remaining number has been multiplied by 5

_____

c) 2 boxes of blue and 2 boxes of red from which 6 counters have been removed from each box and then the remaining number has been doubled

_____

d) 3 boxes of each colour with 8 extra in each box

_____

2 Write an expression for the area and perimeter of each of these rectangles. Use brackets where possible.

a) Rectangle with sides $b+1$ and $3a$

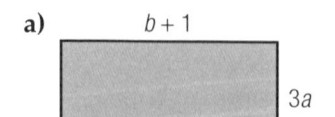

Area = _____

Perimeter = _____

b) Rectangle with sides $2c-1$ and $4b$

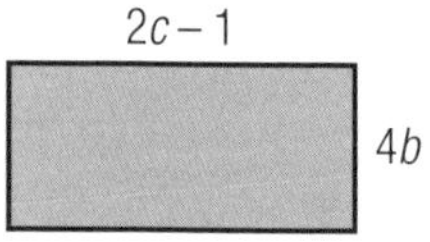

Area = _____

Perimeter = _____

c) Rectangle with sides $5d$ and $3+c$

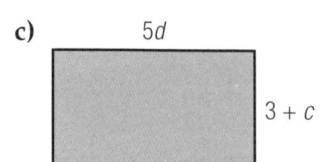

Area = _____

Perimeter = _____

d) Rectangle with sides $3e-1$ and $2d$

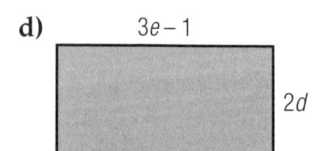

Area = _____

Perimeter = _____

e)   5e

Area = _____

Perimeter = _____

3 Factorise the following expressions fully.

a) $24p - 28q =$ _____

b) $6a - 30b =$ _____

c) $21d - 14e =$ _____

d) $6a + 9b + 12c =$ _____

e) $8a + 2b + 4c =$ _____

f) $6p + 9q + 15r =$ _____

g) $12m + 16n - 36r =$ _____

h) $7a - 14b + 35c =$ _____

i) $24p - 32q - 12r =$ _____

j) $3a - 3b - 3c =$ _____

k) $6a - 12b - 18c =$ _____

l) $7p - 7q + 7r =$ _____

m) $30p - 60q - 15r =$ _____

4 Factorise the following expressions fully.

a) $11a - 11ab =$ _____

b) $4a - 16a - 8ab =$ _____

c) $5ab - 10bc + 15b^2 =$ _____

d) $8ab^2 - 6b^2 =$ _____

e) $a^2 + a =$ _____

f) $b + b^2 =$ _____

g) $b^2 + b^3 =$ _____

h) $a^3 + a^2 + a =$ _____

i) $p^3 + 2p^2 + 3p =$ _____

j) $7m^3 - 9m^2 - 4m =$ _____

k) $56a^2b - 28ab^2 =$ _____

l) $72ab - 36bc + 48bd =$ _____

m) $4a^3b - 6a^3c =$ _____

n) $14m^3n^2 - 21m^2n =$ _____

o) $6a^2b^2 + 12ab =$ _____

p) $3c^2 - 15c^3 =$ _____

q) $5ab - 5ac =$ _____

r) $13b^2c - 26bc^2 =$ _____

CHAPTER 2

5 Factorise the following expressions by grouping.

a) $ac - ad - bc + bd =$ _____

b) $rs - rv - ts + tv =$ _____

c) $wx - vx + vy - wy =$ _____

d) $a^2 + ac + ab + bc =$ _____

e) $pq + p^2 + pr + rq =$ _____

f) $mn + mr + n^2 + rn =$ _____

g) $px + py + rx + ry =$ _____

h) $ab - 2ac - 3bc + 6c^2 =$ _____

i) $ab - a - bd + d =$ _____

Changing the subject of a formula

A **formula** describes a relationship between different variables. The **subject** of the formula is the letter on its own on one side of the equals sign. To make a different letter the subject, rearrange the formula by doing the same to both sides.
Example: Make u the subject of the formula: $v = u + at$.
$v - at = u$ (subtract at from both sides)

Exercise 2.3

Rearrange the following formulae to make the underlined letter the subject.

1 a) $p + \underline{q} = r$

b) $\underline{q} + 2r = s$

$q =$ _____

$q =$ _____

2 a) $2q + \underline{r} = 4p$

b) $3s + \underline{q} = 2p$

$r =$ _____

$q =$ _____

EXPRESSIONS AND FORMULAE

3 a) $pq = r$　　　　　　　　　**b)** $pr = qs$

　　$q =$ _____　　　　　　　　　$r =$ _____

4 a) $pq = r + 3$　　　　　　　**b)** $pr = q - 4$

　　$p =$ _____　　　　　　　　　$r =$ _____

5 a) $m + n = r$　　　　　　　　**b)** $m - n = p$

　　$n =$ _____　　　　　　　　　$n =$ _____

6 a) $2m + n = 3p$　　　　　　　**b)** $3x = 2p + q$

　　$m =$ _____　　　　　　　　　$p =$ _____

7 a) $xy = uv$　　　　　　　　　**b)** $-pq = rs$

　　$x =$ _____　　　　　　　　　$p =$ _____

8 a) $6q = 2p - 5$　　　　　　　**b)** $6q = 2p - 5$

　　$q =$ _____　　　　　　　　　$p =$ _____

9 a) $3x - 7y = 4z$　　　　　　**b)** $3x - 7y = 4z$

　　$z =$ _____　　　　　　　　　$y =$ _____

10 a) $2pr - q = 8$　　　　　　**b)** $2pr - q = 8$

　　$r =$ _____　　　　　　　　　$q =$ _____

13

● CHAPTER 2

11 The distance travelled by an object is expressed in the formula $s = ut + \frac{1}{2}at^2$, where s is the distance, u is the initial (starting) velocity, a is the acceleration and t is the time.

Rearrange the formula to make:

a) u the subject **b)** a the subject.

$u = $ _____ $a = $ _____

12 The formula for calculating the area of a trapezium is $A = \frac{1}{2}(a + b)h$.

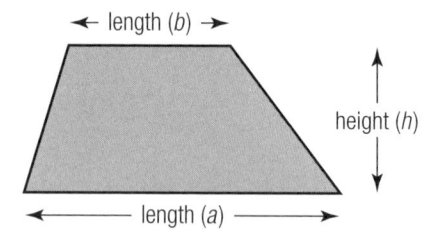

Rearrange the formula to make a the subject.

$a = $ _____

Substitution

We can **substitute** known values for the letters in an expression or formula to find an unknown value.

Example: When $a = 7$ and $b = -3$,

$a + 4b^2$

$= 7 + 4 \times (-3)^2$

$= 7 + 4 \times 9$

$= 7 + 36$

$= 43$

14

Exercise 2.4

1. Calculate the value of each of the following expressions when $a = \frac{1}{2}$, $b = -\frac{1}{2}$, $c = 2$ and $d = -2$. Use a calculator if necessary.

   a) $a + b =$ _____

   b) $a - b =$ _____

   c) $a + b - c =$ _____

   d) $a + b - c - d =$ _____

   e) $6a =$ _____

   f) $4b =$ _____

   g) $6c =$ _____

   h) $2d =$ _____

   i) $a + b + c + d =$ _____

   j) $a - b + c - d =$ _____

   k) $a - b + (c - d) =$ _____

   l) $a - b - (c - d) =$ _____

   m) $(a - b) - (c - d) =$ _____

   n) $a^2 =$ _____

   o) $b^2 =$ _____

   p) $c^2 =$ _____

   q) $d^2 =$ _____

● CHAPTER 2

r) $a^2 - b^2 = $ _____

s) $a^2 - b^2 - c^2 - d^2 = $ _____

2 Given the formula $A = \frac{1}{2}(a + b)h$, find:

  a) the value of A when $a = 12$ cm, $b = 14$ cm and $h = 4$ cm

  _____

  b) the value of h when $A = 60$ cm², $a = 12$ cm and $b = 8$ cm

  _____

  c) the value of b when $A = 15$ cm², $a = 4$ cm and $h = 5$ cm.

  _____

Adding and subtracting algebraic fractions

To add or subtract algebraic fractions, first change them to equivalent fractions with the same denominator if necessary.

Example: $\frac{x}{3} + \frac{y}{4} = \frac{4x}{12} + \frac{3y}{12} = \frac{4x + 3y}{12}$

Exercise 2.5

1 Simplify the following.

  a) $\frac{2}{5} - \frac{1}{10} = $ _____

  b) $\frac{2a}{5} - \frac{a}{10} = $ _____

  c) $\frac{x}{5} + \frac{x}{10} = $ _____

  d) $\frac{3a}{5} - \frac{7a}{25} = $ _____

e) $\frac{3a}{5b} - \frac{7a}{25b} =$ _____

f) $\frac{x}{y} - \frac{2x}{9y} =$ _____

g) $\frac{a}{b} + \frac{c}{3b} - \frac{d}{9b} =$ _____

h) $\frac{ab}{xy} + \frac{2ab}{3xy} =$ _____

i) $\frac{b}{5x} + \frac{b}{20x} - \frac{b}{40x} =$ _____

2 Simplify the following.

a) $\frac{1}{n} + \frac{2}{3n} =$ _____

b) $\frac{m}{n} + \frac{2m}{3n} =$ _____

c) $m + \frac{m}{2} =$ _____

d) $x + \frac{2x}{5} =$ _____

e) $a - \frac{3a}{8} =$ _____

f) $3x + \frac{x}{2} =$ _____

g) $8m - \frac{15m}{2} =$ _____

Teacher comments

3 Shapes and geometric reasoning

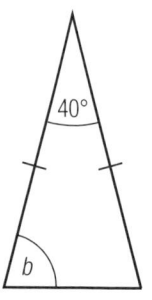

Polygons

A **polygon** is a two-dimensional closed shape with straight sides, for example triangles, quadrilaterals, pentagons and hexagons. The exterior angles of any polygon add up to 360°. The interior and exterior angles at any vertex add up to 180°.

The interior angles of a triangle add up to 180°.

In a **regular** polygon, all the sides are the same length and all the angles are the same size.

- Size of each exterior angle of a regular polygon = $\dfrac{360°}{\text{number of sides}}$

- Size of each interior angle of a regular polygon = $\dfrac{\text{sum of interior angles}}{\text{number of sides}}$

When a diagram shows parallel lines, look for any pairs of **alternate** or **corresponding** angles, which are equal. (Look for 'Z' and 'F' shapes.)

Exercise 3.1

1 Calculate the size of each unknown angle in these diagrams.

a)

$a =$ _____

b)

$b =$ _____

SHAPES AND GEOMETRIC REASONING

c)

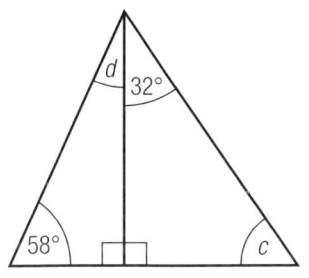

c = _____

d = _____

d)

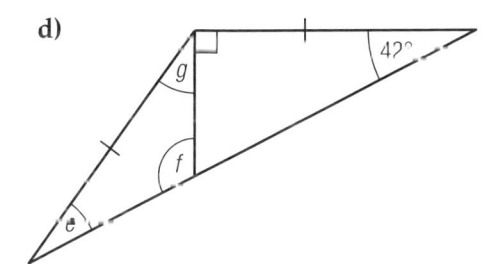

e = _____

f = _____

g = _____

2 Calculate the size of the unknown angles in these polygons.

a)

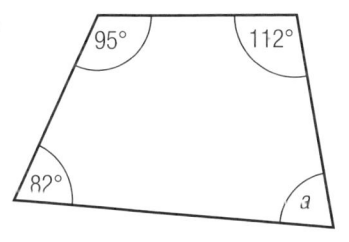

a = _____

b)

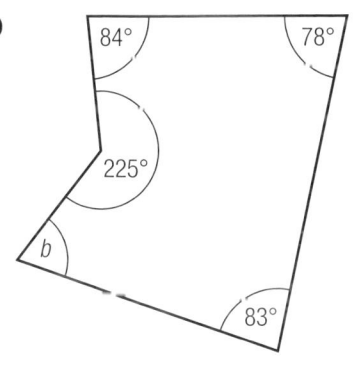

b = _____

● CHAPTER 3

3 A regular decagon has 10 sides as shown.
   Calculate:

   a) the size of each exterior angle

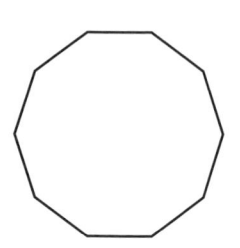

   _____

   b) the size of each interior angle _____

   c) the sum of all the interior angles. _____

4 Calculate the size of each unknown angle in this diagram.

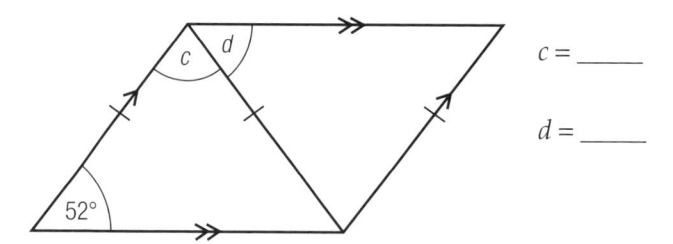

   $c =$ _____

   $d =$ _____

Circle geometry

Exercise 3.2

1 Calculate the size of angle B in this diagram.
   Give reasons for your answer.

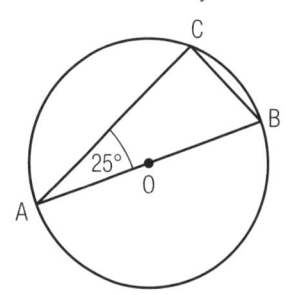

   $B =$ _____

Drawing three-dimensional shapes

A **front elevation** is a two-dimensional view of a three-dimensional object from the front, the **side elevations** are two-dimensional views from the left and right sides, and the **plan** is the view from above. Line up the edges in the different elevations and make sure that the dimensions are consistent. This is easier if a square grid is used.

Views that look three-dimensional can be drawn by using an **isometric grid**. Vertical and horizontal lines on the object are drawn parallel to the three axes of the grid, and shading can be used to make the diagram clearer.

Exercise 3.3

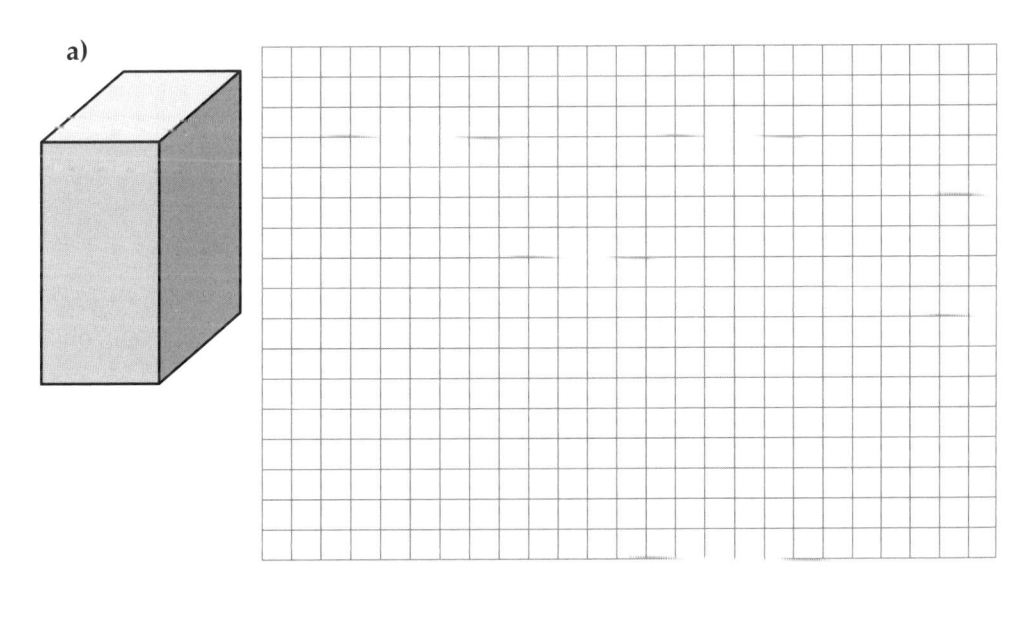

1. For each of the three-dimensional shapes below, draw the possible front, side and plan elevations on the grid provided.

   a)

   b)

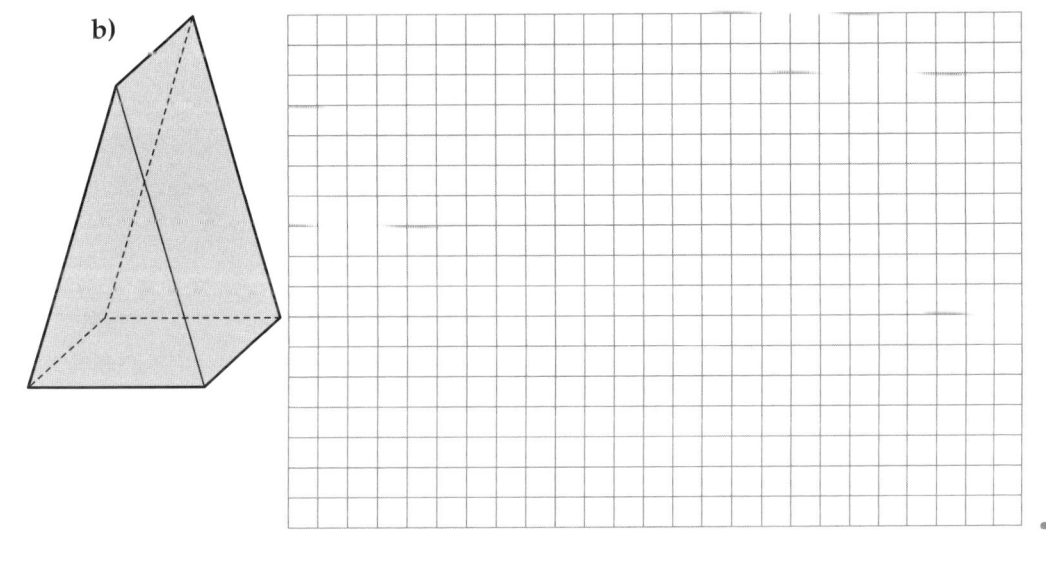

● CHAPTER 3

2   This diagram shows a shape made of small cubes. Draw the front, side and plan elevations on the grid provided. The arrow shows the direction of the front elevation.

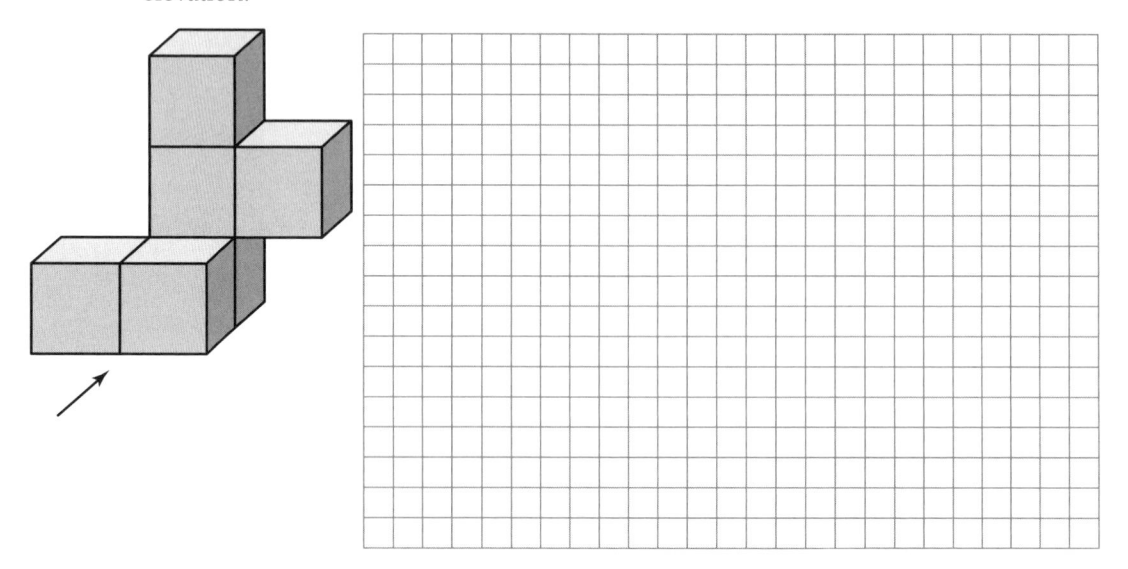

3   For each of the following sets of elevations, sketch the three-dimensional object.

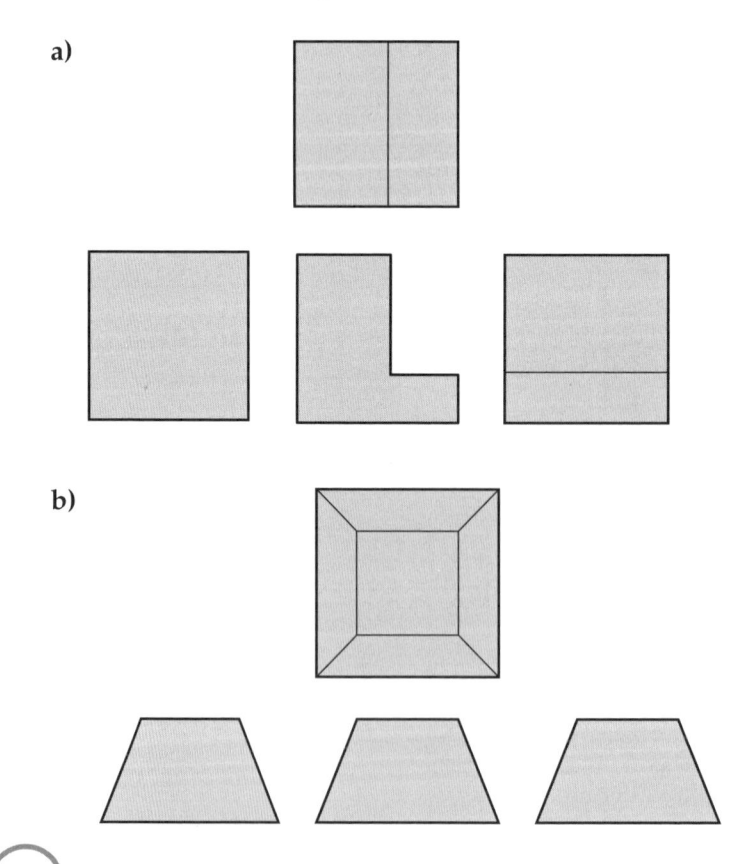

a)

b)

22

4 These diagrams show shapes made of small cubes. Make an isometric drawing of each shape on the grid provided.

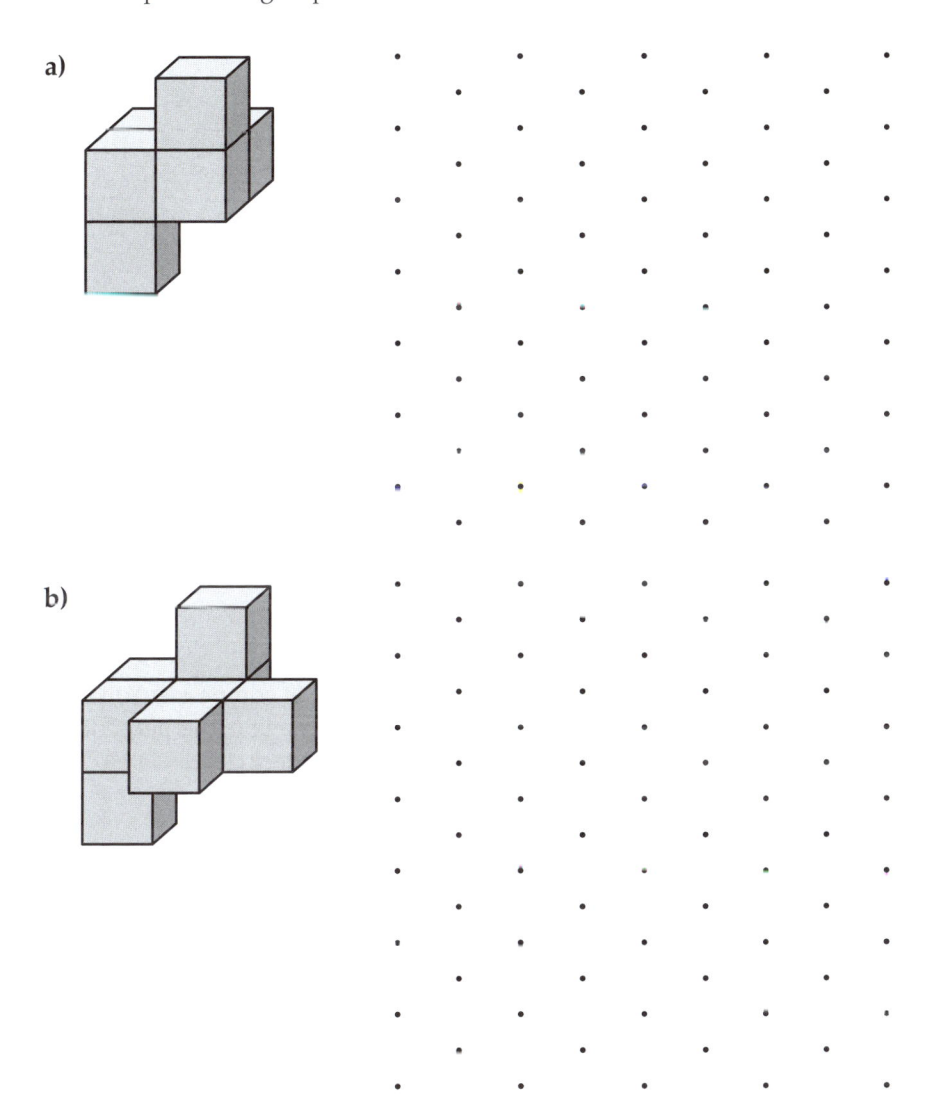

a)

b)

Reflection symmetry in three-dimensional shapes

A **plane of symmetry** divides a three-dimensional shape into two congruent (identical) shapes.

Example: A cuboid has three planes of symmetry.

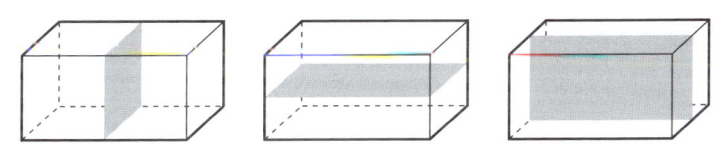

● CHAPTER 3

Exercise 3.4

For each of these diagrams of three-dimensional objects, draw a plane of symmetry and work out how many planes of symmetry the shape has in total.

1 a right-angled triangular prism

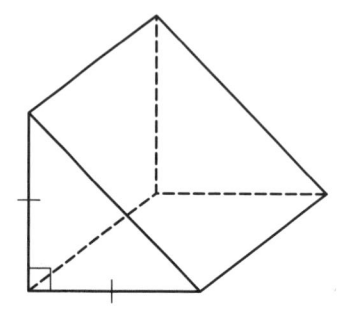

2 a sphere

Number of planes of symmetry = _____

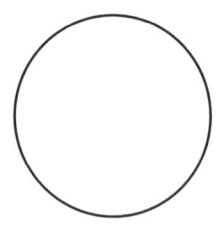

Number of planes of symmetry = _____

Geometric constructions

For geometric constructions, you need a ruler and a pair of compasses. You need to know how to construct the perpendicular from a point to a line, the perpendicular from a point on a line, and a regular polygon inscribed in a circle.

Exercise 3.5

1 Construct a line which passes through the point *P* and is perpendicular to the line *AB*.

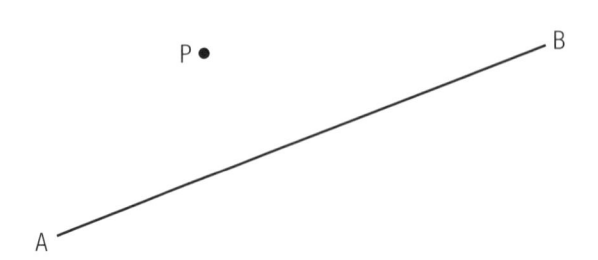

SHAPES AND GEOMETRIC REASONING

2 a) Construct a line which passes through the point M and is perpendicular to the line XY.

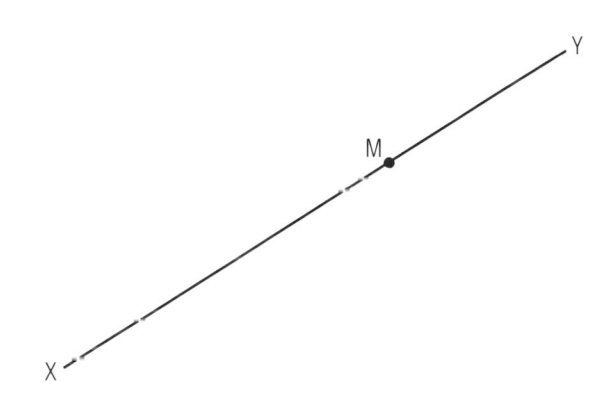

b) Mark a point Z on the perpendicular line so that MZ = 4 cm.

c) Draw the lines XZ and YZ.

d) Calculate the area of the triangle XYZ.

Area of triangle XYZ = _____

● CHAPTER 3

3 Using a ruler and a pair of compasses, construct each of these shapes.

    a) a regular hexagon   **b)** a regular octagon

4 Design a geometric pattern of your own involving at least two squares.

Teacher comments

4 Length, mass and capacity

The metric system

The main units in the metric system are **metres** (for length), **grams** (for masses) and **litres** (for capacities). Different prefixes are used to produce larger or smaller units, in multiples of ten and a thousand.

Example 1: 5 kilograms = 5 × 1000 grams = 5000 g

Example 2: 2 centimetres = 2 × $\frac{1}{100}$ metres = 0.02 m

Exercise 4.1

1 Complete the following.

   a) 600 grams is _____ part of 3.6 kilograms.

   b) 125 kilograms is _____ part of 500 kilograms.

   c) There are _____ litres in 2750 millilitres.

   d) One thousandth of 5 litres is _____.

   e) There are _____ kilograms in 0.375 tonne.

2 a) How much water is left in a 1-litre bottle after 450 ml is drunk?

   b) How much water is left in the bottle after another 355 ml is drunk?

3 A load of cement weighs 2.2 tonnes.
  1625 kg of cement is removed. What weight is left?

CHAPTER 4

4 I am travelling to a city 230 km away.
I take a break half way. How many kilometres do I have left to drive?

_____

5 A motor cyclist travels for five days and covers 1000 km in total.
He covers 123 km, 234 km, 321 km and 109 km on the first four days.
How far does he travel on day 5?

_____

6 a) Name four of the base units from the SI system.

_____   _____   _____   _____

b) Name three units in common use which are not SI units.

_____   _____   _____

7 Write an estimate for each of the following using a sensible unit.

a) the mass of a small block of chocolate _____

b) the length of your foot _____

c) the mass of an average suitcase _____

d) the time it takes to boil an egg _____

e) the number of maths lessons in a school year _____

f) an athlete's speed during a 100 m sprint _____

g) the distance to the nearest city _____

h) the time it takes a ball to fall 100 m _____

8 A pharmacist divides 8 g of a painkiller into 5 mg tablets.
How many tablets are made?

_____

LENGTH, MASS AND CAPACITY

9 I bought 250 g of apples which were priced at $2.48 per kilogram. How much did I pay?

_____

10 a) A piece of rope 15 m long is cut into 1 m 20 cm lengths. How many of these lengths are there?

_____

   b) How long is the piece of rope left over?

_____

11 A golf hole is 342 m from the tee.
A golfer hits their first shot 176 m from the tee, directly towards the hole. How far is the ball from the hole after this shot?

_____

12 A sunflower grows an average of 6 cm per day. How tall is it, in metres, after 31 days?

_____

13 a) My plane takes off at 16 48.
I need to be at the airport 2 hours before the flight. What time should I arrive at the airport?

_____

   b) I am delayed by 48 minutes. What time do I arrive at the airport?

_____

   c) The plane departs 16 minutes late. What time does the plane leave?

_____

Teacher comments

5 Planning and collecting data

You need to **plan** carefully before collecting data, if it is to be reliable and produce useful information.

Exercise 5.1

The questions below are about collecting data to answer this question:

'Are people buying too many cars?'

1 Re-write the question to make it more precise.

   _____

2 What type of primary data could be collected to answer the question?

   _____

3 Give examples of two types of secondary data that could be collected.

   _____

   _____

4 Suggest five questions that could be used in a questionnaire.

   _____

   _____

   _____

   _____

   _____

PLANNING AND COLLECTING DATA

5 Briefly describe how you would collect the primary data. In particular, what sample size would you use and how would you select the sample?

6 How accurate does your primary data need to be? What effect might using a different accuracy have on the data collected?

7 Suggest two ways of improving your data collection.

8 Why should a trial be carried out before the actual data is collected?

Teacher comments

6 Calculations and mental strategies 1

Word problems

Before starting to answer a word problem, decide which mathematical operation (or operations) you need to use.

Exercise 6.1

1 Write a word problem to match each of the following calculations, including the words in brackets in your problem. Then solve the problem.

   a) $20 − $11.40   (total)

   Problem _____

   _____

   Answer = _____

   b) $1 + 3 + 5 + 7 + 9$ (add)

   Problem _____

   _____

   Answer = _____

   c) $16 \times \frac{1}{4}$ (of)

   Problem _____

   _____

   Answer = _____

d) $28 ÷ 7$ (between)

Problem _____

_____

Answer = _____

e) 8×8 (area)

Problem _____

_____

Answer = _____

f) $2(6 + 4)$ (perimeter)

Problem _____

_____

Answer = _____

g) $1.8 - 1.57$ (difference in centimetres)

Problem _____

_____

Answer = _____

h) $270 ÷ 4\frac{1}{2}$ (average speed)

Problem _____

_____

Answer = _____

● CHAPTER 6

2 Write each of the following as a simple maths calculation. Then solve the problem.

a) Find the total of 1.26, 2.14 and 4.05 and divide it by 5.

Calculation _____

Answer = _____

b) What is the sum of the prime numbers between 10 and 20?

Calculation _____

Answer = _____

c) What is the difference between the product of 25 and 4 and the result of dividing 1000 by 8?

Calculation _____

Answer = _____

3 What is the perimeter of a square of side 9 cm?

_____

4 The cost of a meal is $45. A 10% tip is added. What is the total bill?

_____

5 a) A pair of shoes costs $22 and a jacket costs twice as much. What is the total cost?

_____

b) I pay for the shoes and jacket with a $100 note. How much change should I get?

_____

6 What is a half of a half of a half?

_____

7 What is the product of 17 and 3?

_____

8 a) A train travels at an average speed of 160 km/h for $3\frac{1}{2}$ hours. How far does it travel in this time?

_____

b) The train started at 11 15. What time does it arrive?

_____

Dividing by decimals

You can use place value to divide a number by a decimal.
Example: $7.95 \div 3 = 2.65$ so $7.95 \div 0.3 = 2.65 \times 10 = 26.5$ and
$7.95 \div 0.03 = 2.65 \times 100 = 265$

Exercise 6.2

1 Work out the following without using a calculator.

a) (i) $5.61 \div 3 =$ _____

(ii) $5.61 \div 0.3 =$ _____

(iii) $5.61 \div 0.03 =$ _____

b) (i) $6.04 \div 2 =$ _____

(ii) $6.04 \div 0.2 =$ _____

(iii) $6.04 \div 0.02 =$ _____

c) (i) 32.16 ÷ 3 = _____

   (ii) 32.16 ÷ 0.3 = _____

   (iii) 32.16 ÷ 0.03 = _____

d) (i) 198.45 ÷ 5 = _____

   (ii) 198.45 ÷ 0.5 = _____

   (iii) 198.45 ÷ 0.05 = _____

2 Use the fact that 24 243.805 ÷ 98.5 = 246.13 to work out these divisions.

   a) 242.438 05 ÷ 98.5 = _____

   b) 24 243.805 ÷ 0.985 = _____

   c) 24 243.805 ÷ 9.85 = _____

   d) 242.438 05 ÷ 9.85 = _____

   e) 24.243 805 ÷ 0.0985 = _____

3 Use the fact that 7838.428 ÷ 26 = 301.478 to work out these divisions.

   a) 7838.428 ÷ 2.6 = _____

   b) 7.838 428 ÷ 0.26 = _____

   c) 7.838 428 ÷ 0.13 = _____

   d) 7.838 428 ÷ 0.013 = _____

   e) 7.838 428 ÷ 0.52 = _____

Comparison of compound measures – value for money

Quite often a product is offered for sale in different quantities and packages. These different packages are often on sale at different prices. It is therefore important to be able to work out which one offers the **best value for money**.

Exercise 6.3

1 Three taxi firms have different rates per kilometre and for different numbers of passengers:

   Taxi firm A charges $2 per passenger and $0.20 per km.
   Taxi firm B charges $1.50 per passenger and $0.30 per km.
   Taxi firm C charges $0.50 per km for any number of passengers.

   a) Which firm is cheapest if three passengers wish to travel 18 km?

   _____

   b) Which firm is best value for money if one passenger wishes to travel 25 km?

   _____

   c) Give an example of a journey which taxi firm B would provide better value for money than firm C. Show your working, clearly indicating the number of passengers and distance.

Teacher comments

SECTION 2

8 Place value, ordering and rounding

Powers of 10

You can use indices to multiply a number by a power of 10.
Example 1: $6.18 \times 10^3 = 6.18 \times 1000 = 6180$
Example 2: $8.75 \times 10^2 = 8.75 \times 100 = 875$
Example 3: $9.21 \times 10^1 = 9.21 \times 10 = 92.1$
Example 4: $5.73 \times 10^0 = 5.73 \times 1 = 5.73$
Example 5: $5.123 \times 10^{-1} = 5.123 \times 0.1 = 0.5123$
Example 6: $0.45 \times 10^{-2} = 0.45 \times 0.01 = 0.0045$

Dividing by 0.1 or $\frac{1}{10}$ is the same as multiplying by 10 and dividing by 0.01 or $\frac{1}{100}$ is the same as multiplying by 100.
Example 7: $54.2 \div 10^{-1} = 54.2 \div 0.1 = 54.2 \times 10 = 542$
Example 8: $0.987 \div 10^{-2} = 0.987 \div 0.01 = 0.987 \times 100 = 98.7$

Exercise 8.1

Work out the following.

1 a) $7.0 \times 10^3 =$ _____

   b) $7.8 \times 10^3 =$ _____

   c) $7.89 \times 10^3 =$ _____

2 a) $3.0 \times 10^{-1} =$ _____

   b) $3.9 \times 10^{-1} =$ _____

   c) $3.98 \times 10^{-1} =$ _____

PLACE VALUE, ORDERING AND ROUNDING

3 a) $1.0 \times 10^4 =$ _____

   b) $1.4 \times 10^4 =$ _____

   c) $1.42 \times 10^4 =$ _____

4 a) $9.0 \times 10^{-3} =$ _____

   b) $9.5 \times 10^{-3} =$ _____

   c) $9.59 \times 10^{-3} =$ _____

5 a) $7.19 \times 10^2 =$ _____

   b) $7.65 \times 10^{-2} =$ _____

6 a) $8.001 \times 10^3 =$ _____

   b) $5.428 \times 10^{-3} =$ _____

7 a) $5.34 \div 10^2 =$ _____

   b) $5.34 \div 10^1 =$ _____

   c) $5.34 \div 10^0 =$ _____

   d) $5.34 \div 10^{-1} =$ _____

   e) $5.345 \div 10^{-2} =$ _____

   f) $5.456 \div 10^{-3} =$ _____

8 a) $6.678 \div 10^2 =$ _____

   b) $6.567 \div 10^{-1} =$ _____

   c) $6.234 \div 10^{-3} =$ _____

● CHAPTER 8

9 a) $7.02 \div 10^2 =$ _____

   b) $5.92 \div 10^{-2} =$ _____

   c) $3.42 \div 10^{-4} =$ _____

10 a) $8.23 \div 10^1 =$ _____

   b) $3.142 \div 10^0 =$ _____

   c) $3.246 \div 10^{-3} =$ _____

11 a) $0.1 \div 10^2 =$ _____

   b) $0.1 \div 10^{-4} =$ _____

Rounding

To **round** a number to a certain number of significant figures or decimal places, look at the next digit after the one in question. If that digit is 5 or more, round up. If it is 4 or less, round down.

Example 1: 7492 to one significant figure is 7000, and to two significant figures is 7500.

Example 2: 8.785 to one decimal place is 8.8, and to two decimal places is 8.79.

Exercise 8.2

1 Write each of the following numbers to one significant figure.

   a) 6456 _____   b) 885 _____

   c) 73 _____   d) 12 562 _____

   e) 4.78 _____   f) 6.2 _____

   g) 0.34 _____   h) 0.096 _____

PLACE VALUE, ORDERING AND ROUNDING

2 Write each of the following numbers to two significant figures.

a) 6987 _____   b) 64 074 _____

c) 4.572 _____   d) 17.8 _____

e) 59.89 _____   f) 0.800 783 _____

3 Write each of the following numbers to three significant figures.

a) 387 645 _____   b) 476.03 _____

c) 0.228 756 _____   d) 82.5643 _____

e) 82.998 _____

4 Round each of the following numbers to one decimal place.

a) 6.365 _____   b) 4.183 _____

c) 0.1285 _____   d) 8.2472 _____

e) 31.063 _____   f) 0.098 63 _____

5 Round each of the following numbers to two decimal places.

a) 4.3583 _____   b) 5.6719 _____

c) 5.008 03 _____   d) 1.2077 _____

e) 3.899 _____   f) 6.273 _____

6 Round each of the following numbers to three decimal places.

a) 0.004 582 _____   b) 0.007 765 _____

c) 0.099 67 _____   d) 3.870 43 _____

e) 93.004 567 _____

CHAPTER 8

Estimating answers to calculations

An **estimate** is an approximate answer obtained by rounding.

Example 1: Estimate the answer to 48 × 21.
To one significant figure, 48 is 50 and 21 is 20.
So an easy estimate is 50 × 20 = 1000.

Example 2: Estimate the answer to 83 568 ÷ 68.
To two significant figures, 83 568 is 84 000.
To one significant figure, 68 is 70.
So a good estimate would be 84 000 ÷ 70 = 1200.

Exercise 8.3

Estimate the answers to the calculations in questions 1 and 2.
State the number of significant figures in your answers.

1   a) 48 × 18 _____

    b) 69 × 62 _____

    c) 188 × 33 _____

    d) 2.9 × 82 _____

    e) 195 × 37 _____

2   a) 3124 ÷ 21 _____

    b) 5222 ÷ 12 _____

    c) 3.872 ÷ 8.8 _____

    d) 0.414 ÷ 2.1 _____

    e) 0.414 ÷ 0.21 _____

3   Using estimation, decide which of these calculations are definitely wrong.
    Mark them with a tick or a cross.

    a) 48 357 ÷ 291 = 160.8 _____   b) 834 × 0.09 = 988 _____

    c) 5889 ÷ 2.9 = 212 _____   d) 18.3 × 0.02 = 3.66 _____

PLACE VALUE, ORDERING AND ROUNDING

Order of operations

Use **BIDMAS** to remind you of the order of operations in calculations.
Brackets   **I**ndices   **D**ivision/**M**ultiplication   **A**ddition/**S**ubtraction

Exercise 8.4
Work out the following.

1 $2^2 + 3^2 = $ _____

2 $4^2 + 11^2 = $ _____

3 $3^2 - 4^2 = $ _____

4 $3(10^2 - 2^2) = $ _____

5 $2(3^2 + 3^2) - 3(2^2 + 2^2) = $ _____

6 $10(1^2 + 2^2 + 3^2) = $ _____

7 $4 + 4(1^3 + 2^3) - 5(1 + 2 + 3 + 4) = $ _____

8 $2(2^3 + 4^2) + 24 = $ _____

9 $\dfrac{(15^2 - 13^2)}{4} = $ _____

10 $2(5^2 - 2^4) - (3^4 - 3^2) = $ _____

Teacher comments

9 Equations and inequalities

Equations

To **solve** an equation to find the value of an unknown (the variable), rearrange the equation so that the variable is on its own on one side of the equation and everything else is on the other side. Remember to keep the equation balanced by always doing the same to both sides.

Example 1: Solve the equation $6x + 2 = 4x - 8$
$2x + 2 = -8$ (subtract $4x$ from both sides)
$2x = -10$ (subtract 2 from both sides)
$x = -5$ (divide both sides by 2)

Example 2: Solve the equation $4(x + 2) = 3(2x - 2)$
$4x + 8 = 6x - 6$ (expand the brackets)
$8 = 2x - 6$ (subtract $4x$ from both sides)
$14 = 2x$ (add 6 to both sides)
$x = 7$ (divide both sides by 2)

Exercise 9.1

Solve the following linear equations.

1 a) $x + 8 = 12$

   b) $2x - 4 = 8$

   $x =$ _____

   $x =$ _____

   c) $2q + 4 = 4q$

   d) $9q = 15q + 18$

   $q =$ _____

   $q =$ _____

2 a) $4(m - 2) = 24$

   b) $3(3m + 6) = 0$

   $m =$ _____

   $m =$ _____

c) $3(n+2) = 6(n-1)$

d) $4(2n-5) = 2(3n-4)$

$n = $ _____

$n = $ _____

Constructing and solving equations

You can solve a number problem by forming and solving an equation.

Exercise 9.2

Construct an equation from the information given in each question and then solve it.

1. I think of a number and subtract it from 15. The answer is 6.
   What is the number?

   _____

2. I think of a number, divide it by 3 and subtract 12. The answer is 5.
   What is the number?

   _____

3. Add 1 to a number and multiply the result by 2. If this is the same as subtracting 4 and then multiplying the result by 3, what is the number?

   _____

4. Add 3 to a number and divide the result by 2. If this is the same as 6 less than the number, what is the number?

   _____

CHAPTER 9

Simultaneous equations

If there are two equations, both with the same two unknowns, they are called **simultaneous equations**. We need to find the pair of values of the unknowns that satisfies *both* equations. One method for doing this is called the method of **elimination**.

Example 1: Solve these simultaneous equations by elimination.

$2x - y = -2$ equation 1
$5x + y = 16$ equation 2

Add the equations to eliminate y:

$2x - y = -2$
$5x + y = 16$
$\overline{7x = 14}$

Therefore $x = 2$.

To find the value of y, substitute $x = 2$ into equation 2:

$5x + y = 16$
$(5 \times 2) + y = 16$
$10 + y = 16$
$y = 6$

So $x = 2$ and $y = 6$.

Sometimes one of the variables can be eliminated by subtraction rather than addition.

⭐ Sometimes, you cannot eliminate a variable either by addition or by subtraction, and an extra process is needed. One or both equations must be multiplied by a number first.

Example 2: Solve these simultaneous equations by elimination.

$5x - y = 5$ equation 1
$3x + 3y = 21$ equation 2

Multiply equation 1 by 3, to produce two equations containing a $3y$ term:

$15x - 3y = 15$ equation 3
$3x + 3y = 21$ equation 2

Add the equations to eliminate y:

$15x - 3y = 15$
$3x + 3y = 21$
$\overline{ 18x = 36}$

Therefore $x = 2$.

EQUATIONS AND INEQUALITIES

To find the value of y, substitute $x = 2$ into equation 2:

$$3x + 3y = 21$$
$$(3 \times 2) + 3y = 21$$
$$6 + 3y = 21$$
$$3y = 15$$
$$y = 5$$

So $x = 2$ and $y = 5$.

You can solve a number problem involving a pair of unknown numbers by forming and solving a pair of simultaneous equations.

Exercise 9.3

Solve each of the pairs of simultaneous equations in questions 1–8 by elimination.

1 $a + b = 9$
    $4a - b = 6$

$a =$ _____

$b =$ _____

2 $4p + 6q = 22$
    $4p - 2q = 14$

$p =$ _____

$q =$ _____

3 $2x + 3y = 7$
    $-2x + 8y = 26$

$x =$ _____

$y =$ _____

4 $4m - 3n = 9$
    $2m - 3n = -3$

$m =$ _____

$n =$ _____

5 $4a - 2b = 2$
    $5a + b = 20$

$a =$ _____

$b =$ _____

6 $2c + 5d = -13$
    $4c + d = 1$

$c =$ _____

$d =$ _____

● CHAPTER 9

7 $e - 2f = -10$
$8e + 4f = 0$

8 $5g + 2h = 14$
$3g - 4h = -28$

$e =$ _____

$f =$ _____

$g =$ _____

$h =$ _____

9 Form a pair of simultaneous equations for each of these solutions.

  a) $a = 3, b = 4$ **b)** $p = -3, q = 1$

  c) $m = 7, n = -7$ **d)** $r = 5, s = 8$

10 For each of the following problems, form two equations and solve them to find the two numbers.

  a) The sum of two numbers is 11, and their difference is 1.

  b) The sum of two numbers is 13, and their difference is 5.

  c) A number plus 2 times another number is 16. The difference between the two numbers is 4.

Expanding two linear expressions

To expand brackets, multiply all the terms in one set of brackets by the terms in the other set of brackets.

Example: $(x + 3)(x - 4) = x^2 - 4x + 3x - 12 = x^2 - x - 12$

EQUATIONS AND INEQUALITIES

Exercise 9.4

1 For each of the following shapes, write an expression for the area, using brackets. Then expand the brackets and simplify your answer.

a) rectangle with sides $t+11$ and $t+1$

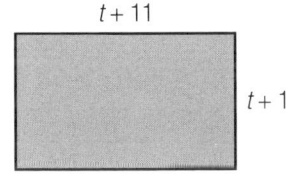

Area = _____

b) right triangle with legs m and $m+6$

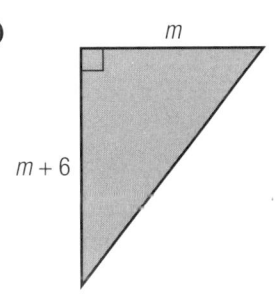

Area = _____

c) L-shape with dimensions $p+7$, p, $p+3$, and 2

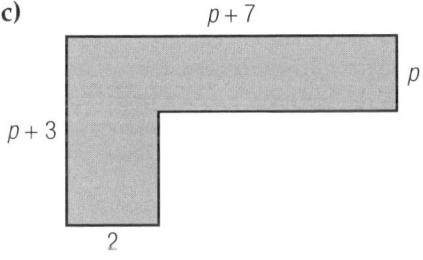

Area = _____

d) triangle with base $h-4$ and height $h+12$

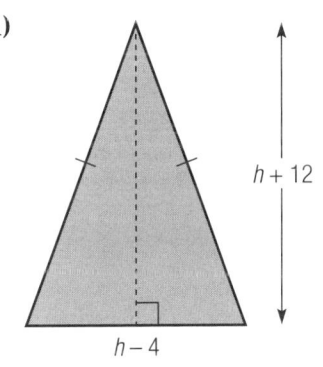

Area = _____

2 Expand the brackets in these expressions and simplify your answer where possible.

a) $(x+5)(x+3) =$ _____

b) $(x-8)(x+4) =$ _____

c) $(x+1)(x+3) =$ _____

d) $(x-7)(x+4) =$ _____

e) $(y+2)(y-1) =$ _____

f) $(y-3)(y+2) =$ _____

Inequalities

An equation shows that one thing is equal to another. An **inequality** shows that one thing is *not* equal to another. The following symbols are used:

- $x < y$ means x is **less than** y.
- $x > y$ means x is **greater than** y.
- $x \leq y$ means x is **less than or equal to** y.
- $x \geq y$ means x is **greater than or equal to** y.
- $x \neq y$ means x is **not equal to** y.

An inequality can be shown on a number line. A closed circle shows that a number *is* included and an open circle shows that a number is *not* included.

Example 1: This number line shows the inequality $x \geq 5$.

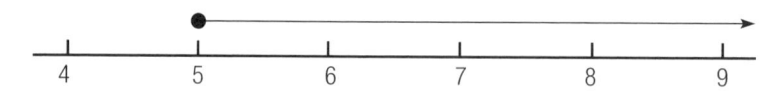

Example 2: This number line shows the inequality $x < 8$.

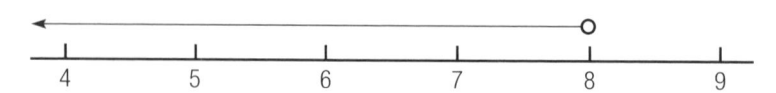

It is possible to combine two inequalities to show that a number lies between two values.

Example 3: $p > 8$ and $p \leq 12$ can be combined as $8 < p \leq 12$, which is shown on a number line like this:

Exercise 9.5

1 Write these inequalities in words.

a) $x < 9$ _____

b) $y \geq 5$ _____

c) $p \neq 9$ _____

d) $12 \leq q$ _____

2 Write one of the signs =, < or > to make each of the following statements true.

a) 8×8 _____ $47 + 19$ b) 15^2 _____ 25×7

c) 15×6 _____ 9×10 d) $1.2\,m$ _____ $130\,cm$

e) 1.05 litres _____ $1500\,ml$ f) $10\,m$ _____ $10\,000\,mm$

g) 2.25 tonnes _____ $2025\,kg$ h) $1\,m^2$ _____ $100\,cm^2$

3 Show each of the following inequalities on the number line provided.

a) $x > 7$

```
  |    |    |    |    |    |
  5    6    7    8    9    10
```

b) $x < 12$

```
  |    |    |    |    |    |
  9   10   11   12   13   14
```

c) $3 > x$

```
  |    |    |    |    |    |
  0    1    2    3    4    5
```

4 Show each of the following inequalities on the number line provided.

a) $5 < a < 9$

```
  |    |    |    |    |    |
  4    5    6    7    8    9
```

CHAPTER 9

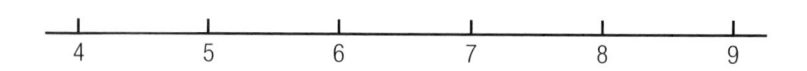

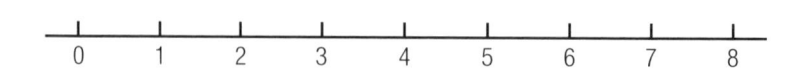

b) $5 \leqslant b < 7$

c) $7 \geqslant c > 1$

d) $-6 < d < -1$

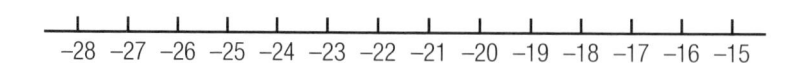

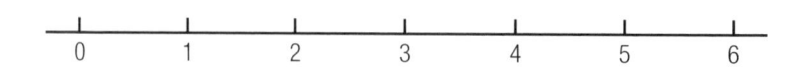

e) $-15 \geqslant e > -28$

f) $6 > g \geqslant 0$

5 Write the inequality shown on each of the following number lines, using the given letter for the unknown.

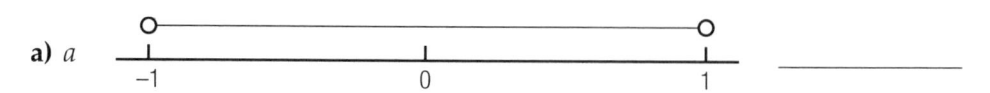

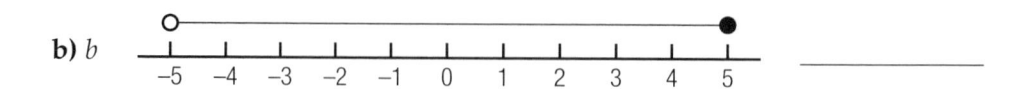

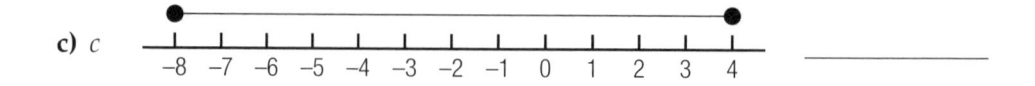

a) a

b) b

c) c

d) d

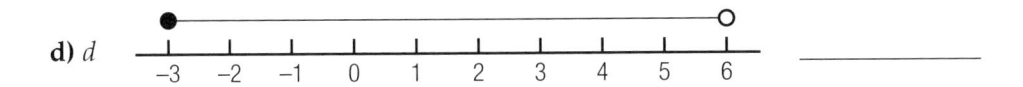

```
●────────────────────────○
-3  -2  -1  0  1  2  3  4  5  6
```

6 Re-write each of the following sentences as an inequality, using the correct inequality sign. Choose a suitable letter to represent the unknown in each case.

a) To go to a particular university a student must be at least 17 years old and not more than 60 years old.

_____

b) To go scuba diving a person must be at least 120 cm tall and not taller than 1.8 m.

_____

c) The temperature in August in Ankara was at least 20 °C at night and reached a maximum of 37 °C during the day.

_____

d) A taxi can carry between 1 and 7 people.

_____

e) Oranges are sorted for market and must be at least 11 cm and less than 15 cm in diameter.

_____

f) A box of tea bags contains more than 36 bags and up to 43 bags.

_____

Solving inequalities

To solve inequalities, use a similar method to the one for solving equations, remembering that you must always do the same to both sides of the inequality.

● CHAPTER 9

Example: Solve the inequality $\quad 5x - 4 > 2x + 2$

$\quad\quad\quad 3x - 4 > 2 \quad$ (subtract $2x$ from both sides)
$\quad\quad\quad 3x > 6 \quad\quad$ (add 4 to both sides)
$\quad\quad\quad x > 2 \quad\quad\;$ (divide both sides by 3)

To solve a combined inequality, first split it into two separate inequalities and then solve each of these separately.

Exercise 9.6

Solve the following inequalities.

1 $x - 3 \leqslant 8$

2 $x + 5 > 1$

3 $2x - 6 \geqslant 1$

4 $-4x + 1 \leqslant x - 9$

5 $\dfrac{2x - 1}{3} > 5$

6 $\dfrac{2 - x}{4} \leqslant -3$

7 $3 \leqslant 2x + 1 < 9$

8 $12 < 3(x + 1) < 24$

Teacher comments

10 Pythagoras' theorem

Pythagoras' theorem states the relationship between the lengths of the three sides of a right-angled triangle:

$a^2 = b^2 + c^2$

In words, the square of the length of the hypotenuse is equal to the sum of the squares of the other two sides.

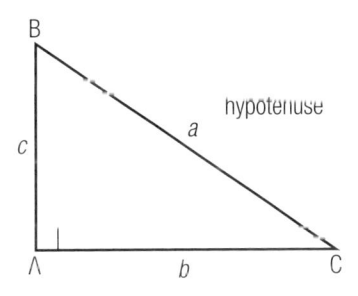

Exercise 10.1

1 Use Pythagoras' theorem to calculate the length of the hypotenuse in each of these right-angled triangles. Give your answers to one decimal place.

a) 7 cm, 5 cm, a

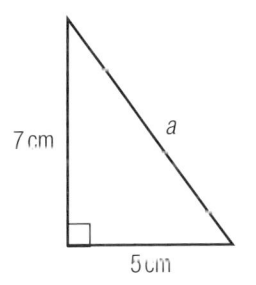

$a = $ _____

b) 9.2 cm, 13.5 cm, b

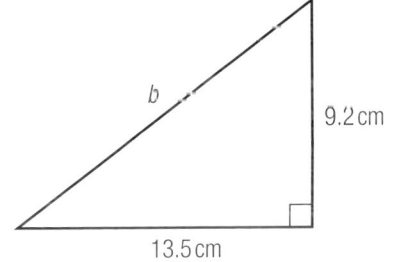

$b = $ _____

● CHAPTER 10

2 Use Pythagoras' theorem to calculate the length of the unknown side in each of these right-angled triangles. Give your answers to one decimal place.

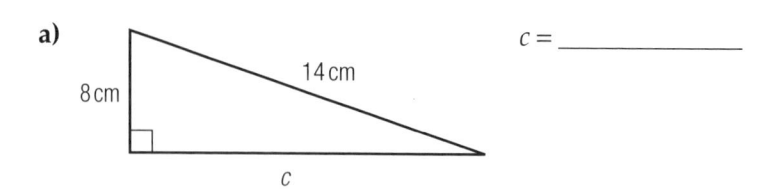

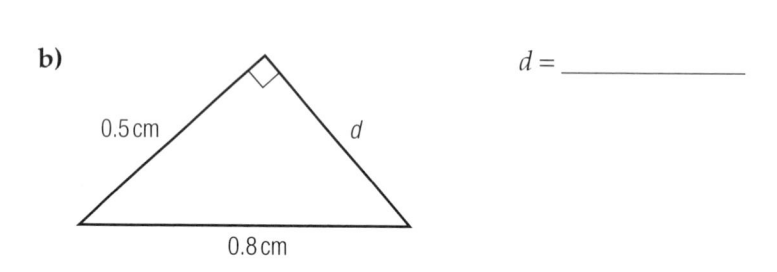

a)

8 cm 14 cm c

$c =$ _____

b)

0.5 cm d 0.8 cm

$d =$ _____

3 Draw a diagram to show the information and use Pythagoras' theorem to solve the problem.

a) A rectangle is 10 cm by 3 cm.
Calculate the length of one of its diagonals.

b) An isosceles triangle has a base of 10 cm and sides of 15 cm.
Calculate its height.

4 Calculate the lengths of the unknown sides in each of these diagrams.

a)

4.1 cm
4.5 cm
6.2 cm
x
y

$x =$ _____

$y =$ _____

b)

2.1 cm
3.9 cm
p
7.2 cm
q

$p =$ _____

$q =$ _____

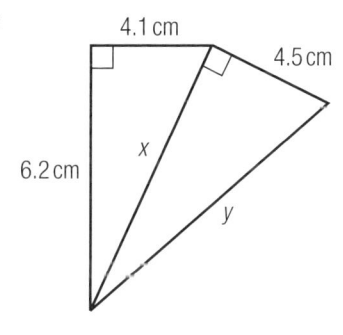

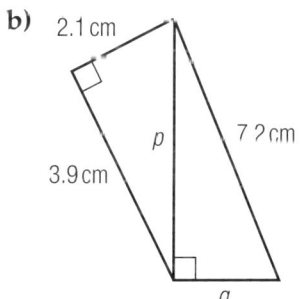

Teacher comments

11 Compound measures and motion

Measures of speed, distance and time

Speed is a compound measure, with units involving distance and time, for example km/h or m/s. The formula linking them can be arranged in three different forms.

- Speed = $\dfrac{\text{distance}}{\text{time}}$
- Distance = speed × time
- Time = $\dfrac{\text{distance}}{\text{speed}}$

The diagram is a helpful way of remembering them.

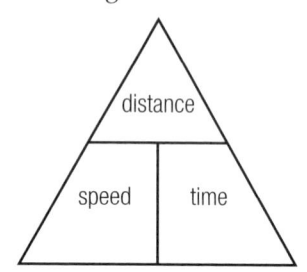

Example 1: Calculate the average speed of a car which makes a journey of 150 km in 2 hours.
Speed = $\dfrac{\text{distance}}{\text{time}} = \dfrac{150}{2} = 75$ km/h.

Example 2: How long does it take a train to travel 480 km at a speed of 160 km/h?
Time = $\dfrac{\text{distance}}{\text{speed}} = \dfrac{480}{160} = 3$ hours.

Exercise 11.1

1 Find the average speed of an object that travels:

   a) 550 km in 11 hours _____

   b) 80 m in 16 seconds _____

   c) 400 m in 20 seconds _____

   d) 150 m in 7.5 seconds _____

   e) 2 km in 30 minutes (give your answer in km/h) _____

f) 25 km in 15 minutes (give your answer in km/h) _____

g) 1 km in 1 minute (give your answer in km/h) _____

h) 2.5 km in 50 seconds (give your answer in km/h) _____

2 How far does an object travel in:

a) 6 hours at 60 km/h _____

b) 15 minutes at 60 km/h _____

c) 20 minutes at 330 km/h _____ _____

d) 10 minutes at 120 km/h _____

e) 8 seconds at 12 m/s _____ _____

f) 45 seconds at 160 m/s _____

g) 10 minutes at 2 m/s _____

h) 2.5 minutes at 20 m/s? _____

3 How long does an object take to travel:

a) 5 km at 30 km/h (give your answer in minutes) _____

b) 5 km at 20 km/h (give your answer in minutes) _____

c) 10 km at 100 km/h (give your answer in minutes) _____

d) 100 m at 10 m/s _____ _____

e) 400 m at 8 m/s _____

f) 800 m at 8 m/s _____

g) 1500 m at 7.5 m/s? _____

CHAPTER 11

4 A bus takes 6 hours 30 minutes to travel a distance of 520 km.
What is its average speed?

_____

5 A boy walks at an average speed of 5.75 km/h.
How long does it take him to travel 28.75 km?

_____

6 A train travels for 4 hours 30 minutes at an average speed of 110 km/h and then for 2 hours 20 minutes at an average speed of 150 km/h.
How far does it travel altogether?

_____

_____

Travel graphs

The motion of an object can be displayed on a **distance–time graph**. If the object is travelling at a constant speed, the distance–time graph is a straight line. The gradient of the line represents the speed.

It can also be shown on a **speed–time graph**. If the object is travelling at a constant speed, the speed–time graph is a *horizontal* line.

Exercise 11.2

1 This graph shows the speed of a car during a journey.

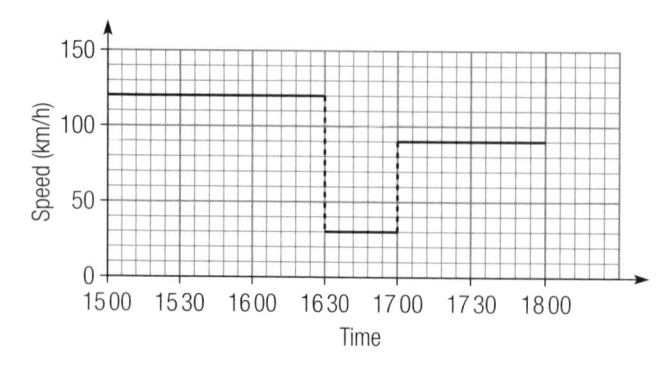

COMPOUND MEASURES AND MOTION

a) What time did the car set off on its journey? _____

b) How fast was the car travelling at 15 30? _____

c) For how long did the car travel at 90 km/h? _____

d) The driver was travelling behind a slow lorry for part of the journey. Between what times was this?

_____

e) How far did the car travel between 15 00 and 16 30?

_____

f) How far did the car travel on the whole journey?

_____

2 Here is a description of a woman's run.

- She sets off at 09 00.
- She runs at a constant speed of 8 km/h for the first hour.
- Then she stops and rests for 15 minutes.
- After her rest she runs at a constant speed of 11 km/h for 20 minutes.

Plot a speed–time graph to show the woman's run.

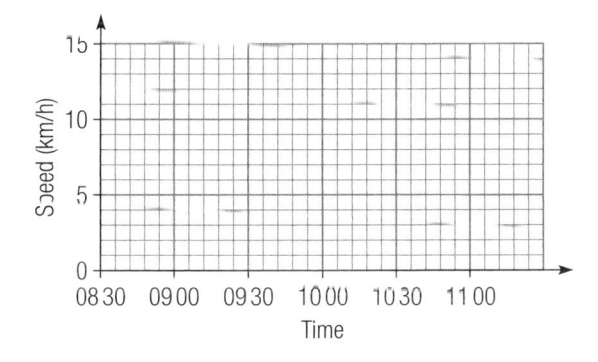

• CHAPTER 11

Graphs in real-life contexts

Graphs can be used to represent many different real-life situations.

Exercise 11.3

1 Two new mobile phones are bought at the same time.
   Phone X costs $150, phone Y costs $100.
   This graph shows the value of the two phones over the next two years.

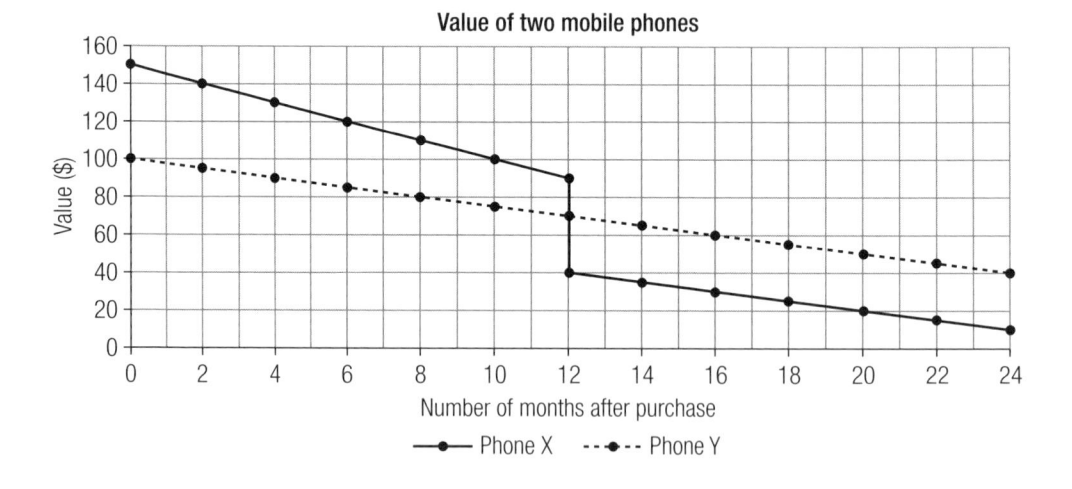

Value of two mobile phones

(Graph: Value ($) vs Number of months after purchase. Phone X (solid line) starts at $150, decreases to $90 at 12 months, drops suddenly to $40, then decreases to about $10 at 24 months. Phone Y (dashed line) starts at $100 and decreases steadily to about $40 at 24 months.)

— Phone X   ---•--- Phone Y

a) Which phone's value falls more quickly during the first year? _____

b) How can you tell this from the shape of the graph?

_____

c) Which phone is better value for money over the two years? _____

d) Give a possible reason for the sudden drop in the value of phone X after one year.

_____

2 Twin babies are washed in a small bath. This graph shows the water level (in centimetres) in the bath over time.

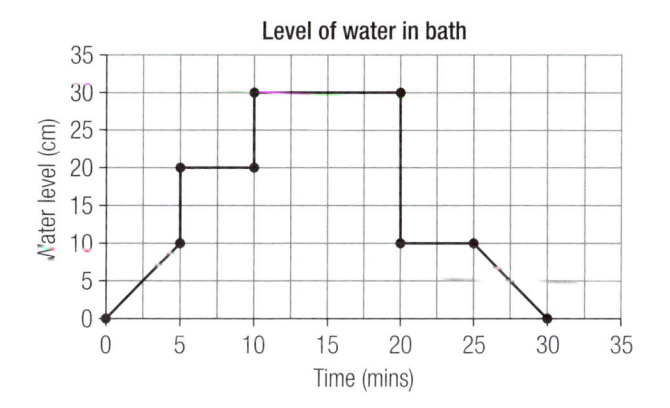

Level of water in bath

Water level (cm) vs *Time (mins)*

a) Give a possible explanation for the shape of the graph during the first 5 minutes.

_____

b) Give an explanation for the shape of the graph at 5 minutes.

_____

c) Give an explanation for the shape of the graph at 10 minutes.

_____

d) For how long were both babies in the bath together?

_____

e) Give an explanation for the shape of the graph at 20 minutes.

_____

Teacher comments

12 Processing and presenting data

Selecting statistics

Many different types of statistical calculation are available, so it is important to select the most appropriate type for the data collected.

Exercise 12.1

In each of the following questions choose the most appropriate calculation to use.

1. These are the numbers of chocolates in 20 similar tubs.

    | 46 | 48 | 49 | 48 | 47 | 50 | 50 | 46 | 60 | 51 |
    | 48 | 50 | 46 | 51 | 60 | 44 | 49 | 52 | 60 | 60 |

    a) What number is the manufacturer most likely to quote as the average number of chocolates in an advertisement?

    _____

    b) Justify your answer to part **a)**.

    _____

    _____

2. Two apple producers each pick 12 apples at random from their orchards and weigh them to the nearest gram. The results are shown below.

    Producer A
    135   136   138   131   140   143   136   137   132   131   133   138

    Producer B
    108   103   134   160   136   153   113   147   169   136   138   133

    They can sell their apples to a supermarket if they are heavier than 132 g on average and are reasonably consistent in weight.

    a) Which producer is more likely to be able to sell their apples to the supermarket?

    _____

b) Justify your answer to part **a)**.

_____

_____

Selecting, drawing and interpreting graphs

It is important to be able to select which type of graph to use to display data as clearly as possible. The decision will depend on the type and quantity of the data.

Exercise 12.2

1. A fishing village keeps a record of the amount (in kilograms) of crabs caught over a number of years. The results are shown in the table.

Year	1990	1991	1992	1993	1994	1995	1996	1997	1998	1999
Amount of crabs (kg)	12 200	11 700	18 100	21 400	13 300	18 000	14 500	73 900	99 500	30 500

Year	2000	2001	2002	2003	2004	2005	2006	2007	2008	2009
Amount of crabs (kg)	30 700	33 900	61 200	89 600	89 100	79 500	42 200	38 900	43 200	77 300

a) What type of graph is the most appropriate to display this data?

_____

● CHAPTER 12

b) Plot the graph you chose in part **a)**.

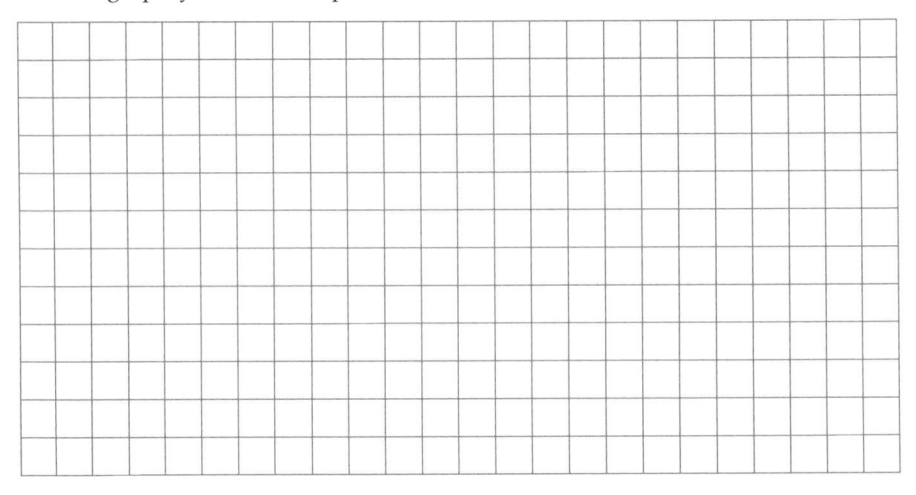

c) Describe the shape of the graph.

_____

_____

d) Suggest a possible reason for the shape of the graph.

_____

_____

2 These are the times (in hours, minutes and seconds) of the first 20 men to complete the Madrid marathon.

02 04 40	02 05 45	02 05 45	02 06 34	02 07 48
02 08 26	02 08 42	02 09 35	02 11 35	02 11 38
02 12 44	02 13 13	02 13 40	02 14 27	02 15 24
02 15 25	02 16 15	02 16 58	02 17 41	02 19 26

a) What type of graph is the most appropriate to display this data?

_____

b) Plot the graph you chose in part **a)**.

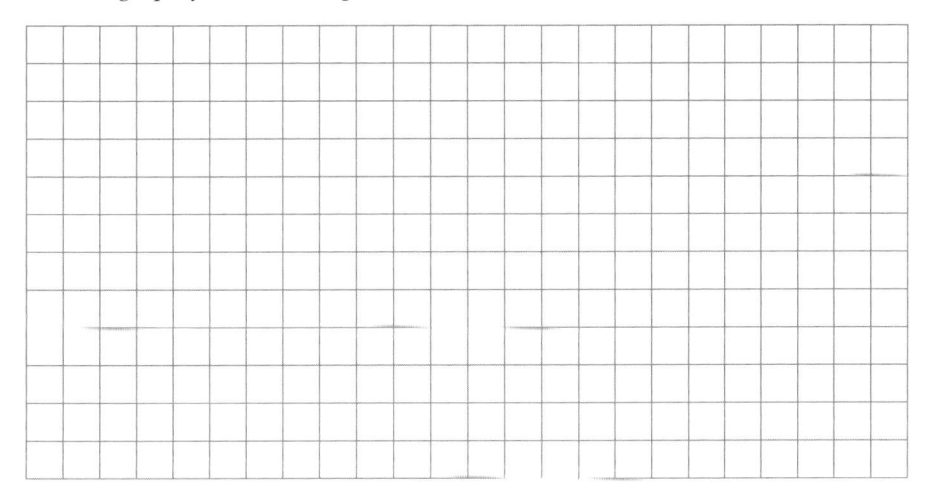

c) Describe the shape of the graph.

_____

Scatter graphs and correlation

Scatter graphs are used to show whether there is a **correlation** (relationship) between two sets of data, which are plotted as coordinate pairs on a single graph. If there is a correlation, a line of best fit can be drawn through the points.

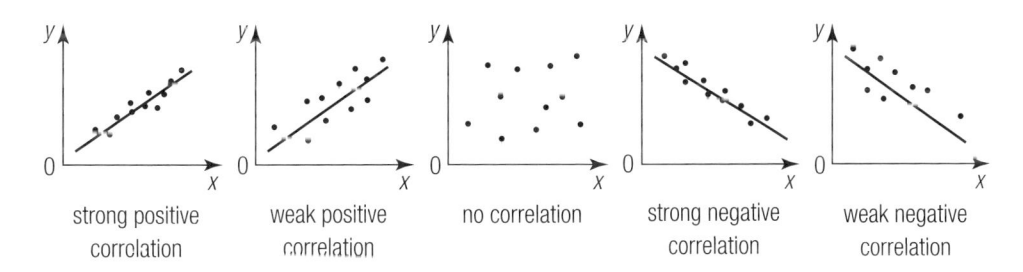

strong positive correlation | weak positive correlation | no correlation | strong negative correlation | weak negative correlation

● CHAPTER 12

Exercise 12.3

1 What type of correlation would you expect (if any) if the following data was collected and plotted on a scatter graph? Give reasons for your answers.

   a) a student's score in a maths exam and their score in a history exam

   _____ because _____

   _____

   b) a student's shoe size and their foot length

   _____ because _____

   _____

   c) the outdoor temperature and the number of scarves sold by a shop

   _____ because _____

   _____

2 This table gives the arm-spans of 16 students in a class and their heights.

Arm-span (cm)	155	160	159	162	161	164	165	170
Height (cm)	160	160	163	156	162	170	168	170

Arm-span (cm)	172	174	175	178	179	182	186	187
Height (cm)	168	182	180	176	181	180	184	189

PROCESSING AND PRESENTING DATA

a) Plot a scatter graph of the data.

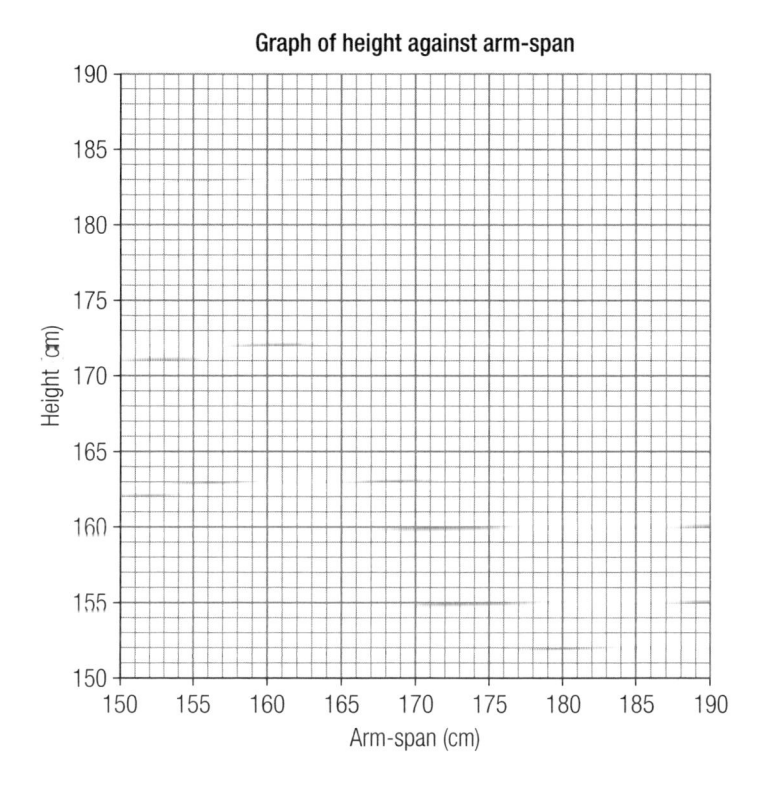

Graph of height against arm-span

[scatter graph with Height (cm) on y-axis from 150 to 190, and Arm-span (cm) on x-axis from 150 to 190]

b) What type of correlation (if any) exists between arm-span and height?

_____

c) Draw a line of best fit on your graph.

d) Use your graph to estimate the height of a student with an arm-span of 171 cm.

_____

● CHAPTER 12

Stem-and-leaf diagrams

Stem-and-leaf diagrams are a special type of bar chart with 'bars' made from the data itself. Two sets of data can be compared by drawing a **back-to-back** stem-and-leaf diagram.

Exercise 12.4

1 A class of 20 students takes a maths test and a geography test. Each test is marked out of 50. The results are shown below.

Maths
21   23   24   34   35   38   39   39   41   41
42   43   43   44   46   46   47   48   49   49

Geography
9   10   14   14   15   17   18   19   20   21
24   25   27   32   37   38   39   41   44   47

a) Display the data on a back-to-back stem-and-leaf diagram.

b) From the diagram, which test appears to have been the harder of the two? Justify your answer.

_____ because _____

_____

Teacher comments

13 Calculations and mental strategies 2

Multiplying by decimals

You can use estimation and place value to help you multiply a number by a decimal.

Example: Work out 7.8×1.9.

Estimate: 7.8×1.9 is about $8 \times 2 = 16$.

Using long multiplication, $78 \times 19 = 1482$.

$7.8 \times 1.9 = (78 \div 10) \times (19 \div 10)$
$= (78 \times 19) \div 100$
$= 1482 \div 100$
$= 14.82$

14.82 is about 16. ✓

Exercise 13.1

For each calculation in questions 1–8, estimate the answer to the calculation. Then work out the exact answer using long multiplication or another method.

1 a) 59×31 b) 5.9×3.1

2 a) 98×49 b) 9.8×4.9

3 a) 82×28 b) 820×2.8

● CHAPTER 13

4 a) 91 × 41 **b)** 0.91 × 4.1

5 a) 26 × 93 **b)** 2600 × 0.93

6 a) 239 × 21 **b)** 23.9 × 2.1

7 a) 1598 × 99 **b)** 15.98 × 9.9

8 a) 232 × 18 **b)** 2320 × 1.8

9 Use the fact that 27 × 27 = 729 to work out these multiplications.

a) 2.7 × 27 = _____

b) 2.7 × 2.7 = _____

10 Use the fact that 73 × 16 = 1168 to work out these multiplications.

a) 73 × 1.6 = _____

b) 73 × 0.16 = _____

CALCULATIONS AND MENTAL STRATEGIES 2

11 Use the fact that 225 × 25 = 5625 to work out these multiplications.

   a) 2.25 × 2.5 = _____

   b) 22.5 × 2500 = _____

12 Use the fact that 12576 × 2 = 25152 to work out these multiplications.

   a) 1.2576 × 0.2 = _____

   b) 12.576 × 200 = _____

13 Use the fact that 847 × 3.6 = 3049.2 to work out these multiplications.

   a) 0.36 × 84.7 = _____

   b) 3.6 × 0.847 = _____

14 Use the fact that 11 × 1111 = 12221 to work out these multiplications.

   a) 1.1 × 1.111 = _____

   b) 0.011 × 111.1 = _____

Teacher comments

SECTION 3

15 Fractions, decimals and percentages

Fractions

To write a fraction in its **simplest form**, divide the numerator and denominator by their highest common factor. The simplification can be done in steps if the highest common factor is not known.

Example 1: Simplify $\frac{27}{63}$

The HCF is 9, so divide top and bottom by 9.

$27 \div 9 = 3$ and $63 \div 9 = 7$. So $\frac{27}{63} = \frac{3}{7}$

To **add** or **subtract** fractions with the same denominator, simply add or subtract the numerators. To add or subtract fractions with different denominators, first change them to equivalent fractions with a common denominator.

Example 2: $\frac{3}{7} + \frac{4}{9} = \frac{27}{63} + \frac{28}{63} = \frac{55}{63}$

To **multiply** fractions, multiply the numerators and multiply the denominators. To keep calculations simple, look for any common factors and cancel these *before* multiplying.

Example 3: $\frac{3}{4} \times \frac{16}{27} = \frac{\cancel{3}}{\cancel{4}} \times \frac{\cancel{16}^{4}}{\cancel{27}_{9}} = \frac{4}{9}$

To **divide** by a fraction, remember that multiplying by $\frac{1}{7}$ is the same as dividing by 7. Similarly, dividing by $\frac{3}{2}$ is the same as multiplying by $\frac{2}{3}$.

Example 4: $\frac{4}{5} \div \frac{2}{25} = \frac{4}{5} \times \frac{25}{2} = \frac{\cancel{4}^{2}}{\cancel{5}} \times \frac{\cancel{25}^{5}}{\cancel{2}} = 10$

Exercise 15.1

1. Write each of the following fractions in its simplest form.

   a) $\frac{54}{72} =$

   b) $\frac{36}{108} =$

FRACTIONS, DECIMALS AND PERCENTAGES

c) $\frac{77}{121} =$

d) $\frac{26}{65} =$

e) $\frac{19}{24} =$

f) $\frac{104}{312} =$

2 Do the following calculations.

a) $\frac{3}{11} + \frac{4}{11} =$

b) $\frac{14}{37} - \frac{12}{37} =$

c) $\frac{7}{23} + \frac{13}{23} - \frac{15}{23} =$

d) $\frac{16}{29} + \frac{17}{29} - \frac{23}{29} =$

3 Do the following calculations. Give your answers in their simplest form.

a) $\frac{1}{3} - \frac{1}{6} =$

b) $\frac{1}{2} - \frac{1}{10} =$

c) $\frac{1}{2} - \frac{1}{5} =$

d) $\frac{1}{2}+\frac{1}{8}-\frac{1}{4}=$

e) $\frac{5}{18}-\frac{5}{9}+\frac{5}{6}=$

4 Do the following calculations. Cancel any common factors first and give your answers in their simplest form.

a) $\frac{1}{6}\times\frac{3}{8}=$

b) $\frac{7}{15}\times\frac{5}{14}=$

c) $\frac{1}{8}\times 1\frac{1}{4}\times\frac{1}{5}=$

5 Do the following calculations, showing your working clearly. Cancel any common factors first and give your answers in their simplest form.

a) $\frac{5}{9}\div\frac{1}{3}=$

b) $\frac{3}{14}\div\frac{5}{21}\times\frac{1}{2}=$

c) $\frac{3}{16}\div\frac{3}{4}\times\frac{1}{3}=$

d) $\frac{11}{42} \div \frac{7}{21} =$

e) $\frac{9}{16} \div \frac{3}{8} \div \frac{1}{8} =$

f) $\frac{1}{2} : \frac{1}{3} : \frac{15}{16} =$

Finance – discount, profit and loss, interest and tax

The formula for calculating **simple interest** is

$$I = \frac{ptr}{100}$$

where I is the interest paid by the bank
p is the principal
t is the time in years
r is the rate per year (percentage).

Example 1: Find the simple interest earned on a deposit of $200 saved for 5 years at a rate of 3% per year.

$$I = \frac{ptr}{100} = \frac{200 \times 5 \times 3}{100} = 30$$

So the simple interest earned is $30.

Example 2: How long will it take a principal of $200 to earn simple interest of $160 at 4% per year?

$$I = \frac{ptr}{100}$$

$$160 = \frac{200 \times t \times 4}{100}$$

$$160 = 8t$$

$$20 = t$$

Therefore it will take 20 years.

A **profit** is made when the selling price of an item is more than it cost to make (its cost price). A **loss** is made when the selling price is less than the cost price. The percentage profit or loss is given by the formula:

$$\text{Percentage profit or loss} = \frac{\text{profit or loss}}{\text{cost price}} \times 100$$

Example 3: A market trader buys 100 oranges for $25 and sells all of them for 30 cents each. What is the profit or loss?

Cost price = $25
Selling price = 100 × 30 cents = $30
Profit = selling price − cost price = $30 − $25 = $5
His profit is $5.

Example 4: A house is bought for $200 000 and sold the following year for $216 000. What is the percentage profit?

Profit = $216 000 − $200 000 = $16 000

$$\text{Percentage profit} = \frac{\text{profit}}{\text{cost price}} = \frac{16\,000}{200\,000} \times 100 = 8$$

The percentage profit was 8%.

Exercise 15.2

All the rates of interest in this exercise are annual rates of simple interest.

1 Calculate the interest paid in each of the following cases.

   a) principal $1500   time 5 years   rate 3%

   b) principal $250   time 10 years   rate 2.5%

2 Calculate how many years it will take to earn the given amount of interest in each of the following cases.

   a) principal $2000   rate 10%   interest $500

   b) principal $75 000   rate 4%   interest $1500

3 What rate of interest per year will earn the given amount of interest in each of the following cases?

   a) principal $8000	time 5 years	interest $800

   b) principal $2000	time 3 years	interest $360

4 What principal will earn the given amount of interest in each of the following cases?

   a) interest $112	time 2 years	rate 3.5%

   b) interest $2240	time 3.5 years	rate 20%

5 For each of the following items:
   (i) work out the percentage profit or loss
   (ii) work out how much tax is added if a 15% tax is added to the selling price.

   a) toaster	cost price $25	selling price $30

   b) coffee maker	cost price $20	selling price $30

   c) food processor	cost price $150	selling price $200

CHAPTER 15

d) fridge   cost price $400   selling price $480

e) freezer   cost price $300   selling price $360

f) Mazda   cost price $12 000   selling price $3000

g) apartment   cost price $250 000   selling price $275 000

h) apartment   cost price $500 000   selling price $400 000

i) Mercedes   cost price $40 000   selling price $5000

Teacher comments

16 Sequences

Sequences

A **sequence** is an ordered set of numbers. Each number in the sequence is called a **term**. The **term-to-term rule** describes how to get from one term to the next.
In an **arithmetic** sequence, there is a constant difference between successive terms.
In a **geometric** sequence, there is a constant ratio between successive terms.

Example 1: The term-to-term rule for the sequence −5, −3, −1, +1, ... is +2.
It is an arithmetic sequence with common difference +2.

Example 2: The term-to-term rule for the sequence 1, 2, 4, 8, ... is ×2.
It is a geometric sequence with common ratio ×2.

Exercise 16.1

For each of the sequences below, write down the next two terms, describe the term-to-term rule, and write if it is arithmetic (A), geometric (G) or neither (N).

1   3   6   9   12   15   ___ ___   Rule: _____   Type: ___

2   −1   −3   −5   −7   −9   ___ ___   Rule: _____   Type: ___

3   32   16   8   4   2   ___ ___   Rule: _____   Type: ___

4   1.1   0.9   0.7   0.5   0.3   ___ ___   Rule: _____   Type: ___

5   1   3   9   27   81   ___ ___   Rule: _____   Type: ___

6   2   5   10   17   26   ___ ___   Rule: _____   Type: ___

7   1000   100   10   1   0.1   ___ ___   Rule: _____   Type: ___

8   1   1   2   3   5   ___ ___   Rule: _____   Type: ___

9   1   −9   −19   −29   ___ ___   Rule: _____   Type: ___

10   ?   ?   ?   −2   2   ___ ___   Rule: _____   Type: ___

CHAPTER 16

Position-to-term rules – the nth term

The **position-to-term rule** describes how to calculate the value of a term from its position in the sequence. We can use algebra to write it as a formula for the value of the **nth term**, where n is the term's position.

For an arithmetic sequence, the nth term is

$$t_n = t_1 + (n-1)d$$

where t_1 is the value of the first term and d is the constant difference.

Example: The table shows the first few terms in a sequence.

Find the 21st term and a simplified formula for the nth term.
$t_n = t_1 + (n-1)d$
For this sequence, $t_1 = 11$ and $d = 4$
The 21st term is $t_{21} = 11 + (21-1) \times 4 = 11 + (20 \times 4) = 91$
The nth term is $t_n = 11 + (n-1) \times 4 = 11 + 4n - 4 = 4n + 7$

Position, n	1	2	3	4	5
Term	11	15	19	23	27

Exercise 16.2

For each of the sequences below, derive a simplified formula for the nth term of the sequence and use the position-to-term rule to find the value of the term shown.

1   32   34   36   38   40   ...

nth term = _____

$t_{10} =$ _____

2   105   110   115   120   ...

nth term = _____

$t_{101} =$ _____

3 29 31 33 35 37 ...

    nth term = _____

    t_{10} = _____

4 7 4 1 −2 −5 ...

    nth term = _____

    t_{12} = _____

5 11 1 −9 −19 ...

    nth term = _____

    t_{11} = _____

6 −66 −63 −60 −57 −54 ...

    nth term = _____

    t_{16} = _____

● CHAPTER 16

7 36 42 48 54 60 ...

nth term = _____

t_{21} = _____

8 −8 −4 0 4 8 12 ...

nth term = _____

t_{101} = _____

9 −3 −8 −13 −18 ...

nth term = _____

t_{51} = _____

10 2 2.5 3 3.5 4 ...

nth term = _____

t_{17} = _____

Teacher comments

17 Position and movement

Tessellations

Shapes **tessellate** if they fit together without leaving any gaps. Here are two examples of shapes tessellating with themselves.

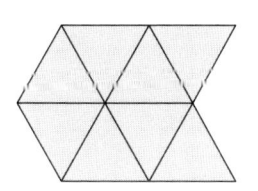

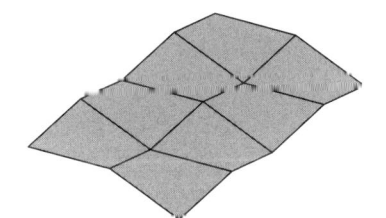

Not all shapes tessellate with themselves. For example, regular pentagons do not tessellate.

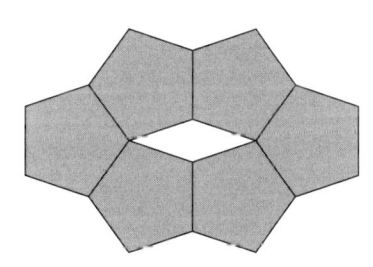

Exercise 17.1

1. Draw a tessellating pattern of scalene triangles on the grid provided.

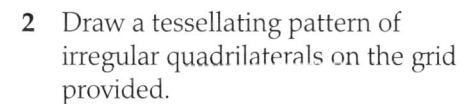

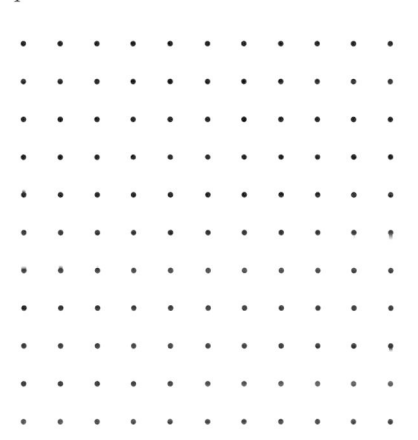

2. Draw a tessellating pattern of irregular quadrilaterals on the grid provided.

● CHAPTER 17

3 This diagram shows isosceles triangles tessellating.
One of the angles is 70°.

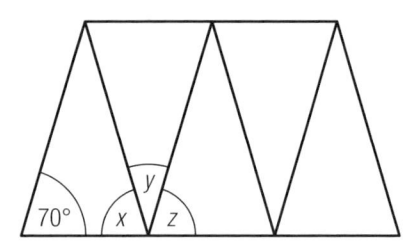

a) What is the size of angle x? _____

b) What is the size of angle y? _____

c) What is the size of angle z? _____

d) Explain why the sizes of angles x, y and z prove that the triangles will continue to tessellate in all directions.

_____

_____

4 This diagram shows the sizes of the angles of an irregular quadrilateral.

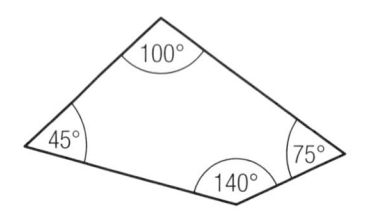

a) In this diagram, an identical quadrilateral has been rotated by 180° and placed next to the original one.
Label each of the angles in this quadrilateral with its size.

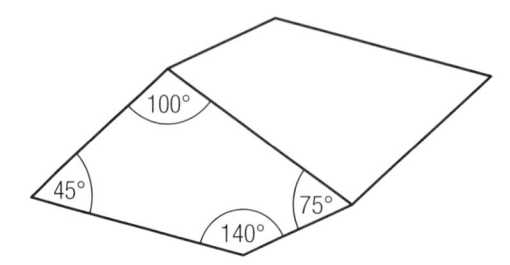

POSITION AND MOVEMENT

b) Here, two extra identical quadrilaterals have been added to the pattern. Label each of the unmarked angles that meet at the centre with its size.

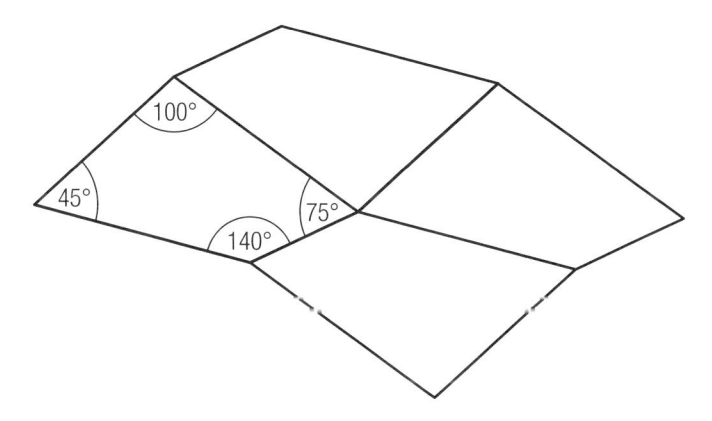

c) Use your answer to part **b)** to explain why the shape tessellates.

_____

_____

5 **a)** Write down the size of an interior angle of a regular hexagon. _____

b) Use your answer to part **a)** to explain why regular hexagons tessellate.

_____

_____

c) Draw a tessellating pattern of regular hexagons on the grid provided.

● CHAPTER 17

6 a) A decagon is a polygon with ten sides.
Calculate the size of an interior angle of a regular decagon.

_____

b) This diagram shows why regular decagons do not tessellate by themselves.
Write down the name of the shape needed to form a tessellating pattern with the decagons.

_____

c) What are the sizes of the two different angles in the central shape?

_____   _____

Transformations

Enlargement

Transformations include reflection, rotation and translation. In these, the image is **congruent** to the original object.

In an **enlargement**, the image is mathematically **similar** to the original object. The angles remain unchanged but the object lengths are all multiplied by the same amount (the **scale factor of enlargement**) to give the image lengths.

To fully describe a transformation:

- for a reflection, give the equation of the mirror line
- for a rotation, give the angle, the direction and the coordinates of the centre of rotation
- for a translation, give the vector
- for an enlargement, give the scale factor of enlargement and the coordinates of the centre of enlargement.

When transformations are combined, the types of the transformations and the order in which they are carried out affect where the image appears.

Exercise 17.2

1. Enlarge each of the objects below by the given scale factor and from the centre of enlargement O.

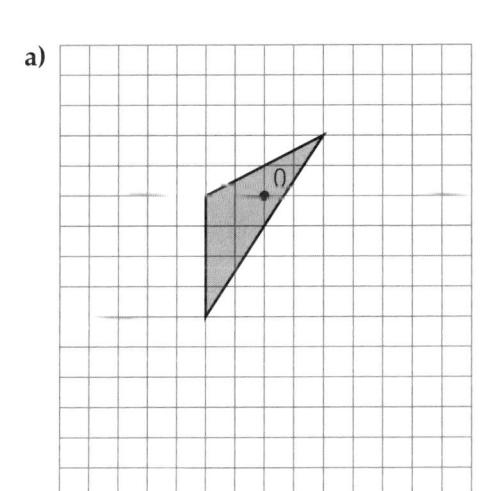

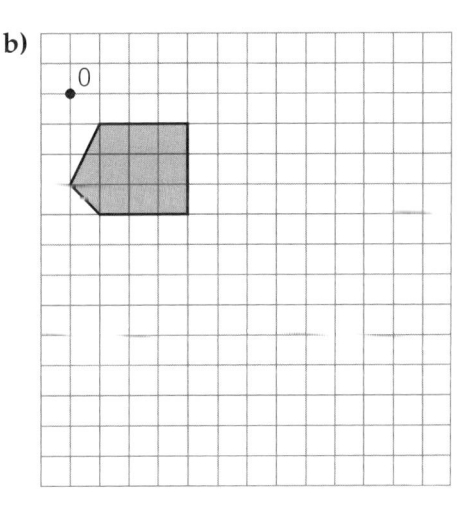

   a)

   Scale factor of enlargement 2

   b)

   Scale factor of enlargement 3

2. In each of the diagrams below, the larger shape is an enlargement of the smaller one from the centre of enlargement O. Work out the scale factor of enlargement in each case.

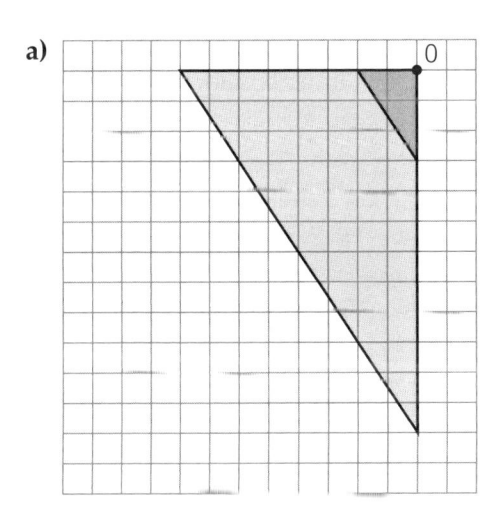

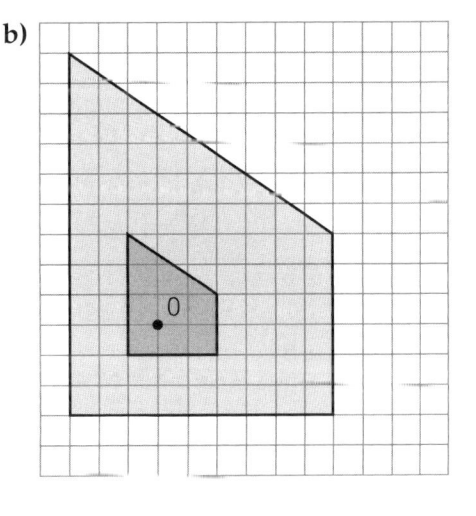

   a)

   Scale factor = _____

   b)

   Scale factor = _____

● CHAPTER 17

3 In each of the diagrams below, the object P undergoes two transformations. The first transformation maps P on to an image Q, the second maps Q on to an image R. Draw each of the images Q and R, labelling them clearly.

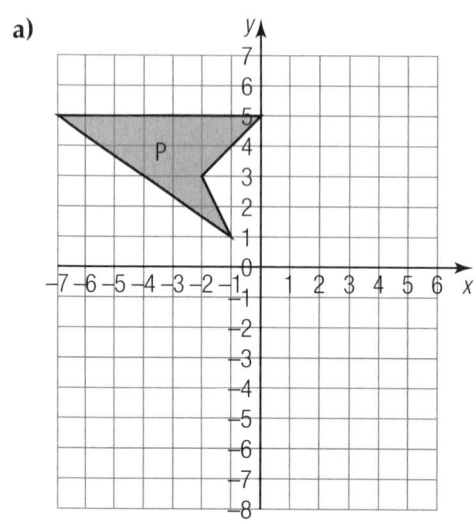

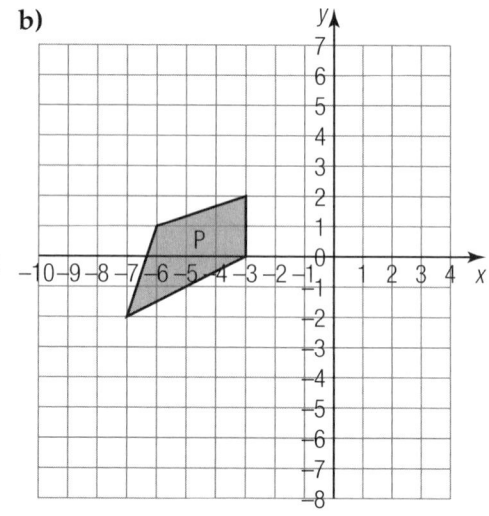

a)

b)

- A reflection in the line $y = 1$
- A rotation by 90° anti-clockwise about $(-1, 1)$

- A translation of $\begin{pmatrix} 4 \\ 1 \end{pmatrix}$
- An enlargement with scale factor 2 and centre of enlargement at $(1, 4)$

4 The object X undergoes two transformations. The first transformation maps X on to the image Y, the second maps Y on to the image Z. Fully describe each transformation, marking any mirror line, centre of rotation or centre of enlargement on the diagram.

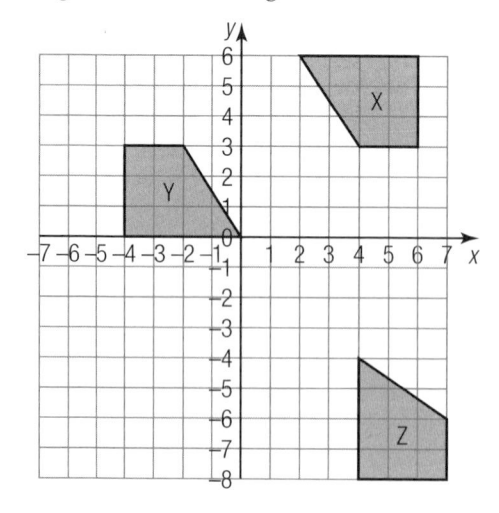

Transformation from X to Y:

_____

_____

Transformation from Y to Z:

_____

_____

Teacher comments

18 Area and volume

Metric units of area

To convert:

from cm² to mm², multiply by 100
from mm² to cm², divide by 100
from m² to cm², multiply by 10 000
from cm² to m², divide by 10 000.

Example 1: 320 cm² = 320 × 100 = 32 000 mm²
Example 2: 6500 mm² = 6500 ÷ 100 = 65 cm²

Exercise 18.1

1 Convert each of these areas into the units shown.

a) 25 000 mm² = _____ cm²   b) 8 cm² = _____ mm²

c) 3 m² = _____ cm²   d) 40 000 cm² = _____ m²

e) 2.8 m² = _____ cm²   f) 25 000 cm² = _____ m²

g) 8000 cm² = _____ m²   h) 0.35 m² = _____ cm²

i) 500 cm² = _____ m²   j) 0.03 m² = _____ cm²

2 Find the area of each of these. Give your answers in the units shown in brackets.

a) a tile measuring 20 cm by 20 cm   (cm²)

_____

b) the tile in part **a)**   (mm²)

_____

● CHAPTER 18

c) a card measuring 50 mm by 100 mm (mm²)

_____

d) the card in part **c)** (cm²)

_____

e) a book cover of area 450 cm² (mm²)

_____

f) a stamp of area 320 mm² (cm²)

_____

g) a small tile measuring 2.5 cm by 2.5 cm (cm²)

_____

h) the tile in part **g)** (mm²)

_____

i) a board measuring 1 m by 1.5 m (cm²)

_____

j) a gate measuring 250 cm by 400 cm (m²)

_____

AREA AND VOLUME

Metric units of volume and capacity

To convert:

from cm³ to mm³, multiply by 1000
from mm³ to cm³, divide by 1000
from m³ to cm³, multiply by 1 000 000
from cm³ to m³, divide by 1 000 000.

Example: 4.8 cm³ = 4.8 × 1000 = 4800 mm³

Litres and millilitres are metric units of liquid volume and capacity.
1 m*l* has a volume of 1 cm³ and 1 litre is 1000 m*l*.

Exercise 18.2

1 Convert each of these volumes into the units shown.

   a) 6.2 cm³ = _____ mm³   b) 8300 mm³ = _____ cm³

2 Find the volume of each of these. Give your answers in the units shown in brackets.

   a) a box measuring 2 cm by 2 cm by 2 cm   (cm³)

   _____

   b) the box in part a)   (mm³)

   _____

3 What is the capacity of the box in question 2? Give your answer in millilitres.

   _____

4 The diagram shows the net of a cube of side length 2.2 cm.

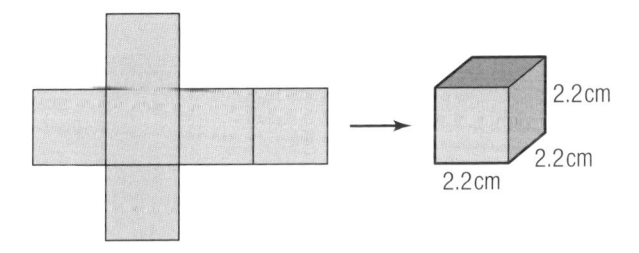

● CHAPTER 18

a) Find the area of the net:

(i) in mm² _____

(ii) in cm² _____

b) If the net is folded into the cube, find the volume of the cube:

(i) in mm³ _____

(ii) in cm³ _____

c) What is the capacity of the cube in part **b)** in millilitres?

_____

Land area

The metric unit of land area is the hectare.

1 hectare = 10 000 m²

Example 1: 4.25 hectares = 4.25 × 10 000 = 42 500 m²

Example 2: 250 000 m² = 250 000 ÷ 10 000 = 25 hectares

Exercise 18.3

1 a) A football training area is a rectangle 200 m by 100 m. What is its area in hectares?

_____

b) The training area is divided into eight equal squares. What is the area of each square, in square metres?

_____

AREA AND VOLUME

2. A hotel is built on a plot of land of area 0.92 hectares.
   What is its area in square metres?

   _____

3. **a)** A new housing estate is built on a square piece of land of side length 1.2 km.
   What is its area in hectares?

   _____

   b) 288 houses are built on the site.
   What is the average (mean) size of each building plot, in square metres?

   _____

4. **a)** A golf course covers an area of 36 hectares.
   What is its area in square metres?

   _____

   b) 25% of the area of the golf course is used for cut grass.
   What area is used for cut grass, in hectares?

   _____

   c) There are 18 holes on the course.
   What is the average (mean) area of cut grass for each hole, in square metres?

   _____

Teacher comments

19 Interpreting and discussing results

It is important to be able to interpret results which have been presented in different forms, for example as line graphs or pie charts.

Exercise 19.1

1. A group of students think that someone's foot length (in centimetres) will give a good indication of their height (in centimetres). They collect data to test this theory. This graph shows their results.

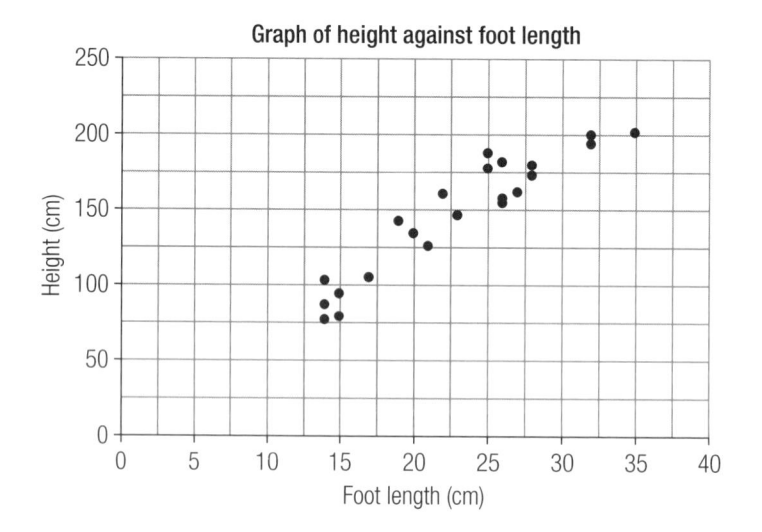

Graph of height against foot length

a) What type of graph has been plotted? _____

b) Does the graph support the students' theory that there is a correlation between someone's foot length and their height?

_____ because _____

_____

INTERPRETING AND DISCUSSING RESULTS

c) One student thinks that if a line of best fit is drawn, it can be extended to make predictions about people's height and foot length. Explain why extending the line of best fit in this case is not likely to produce sensible results.

_____

_____

2 This graph shows the distribution of the lengths of snakes of two different species, P and Q. The mean length of a snake from species P is M_1; the mean length of a snake from species Q is M_2.

a) Which species has the larger mean length? _____

b) Which species has lengths which are within a smaller range? Justify your answer.

_____ because _____

_____

c) Another snake is caught and its length is measured. Its length is x cm (shown on the graph). Is this snake more likely to be from species P or species Q? Justify your answer.

_____ because _____

_____

Teacher comments

97

20 Calculations and mental strategies 3

Factors

The **highest common factor** (HCF) of two or more numbers is the largest number which is a factor of all of them.

Exercise 20.1

Find the highest common factor of the following numbers.

1 16, 36, 44 _____   2 27, 72, 45 _____

3 65, 52, 39 _____   4 48, 72, 24 _____

5 13, 19, 23 _____

Fractions

You can use factors to simplify fractions and calculations with fractions. Remember to look for any common factors which cancel.

Example 1: Simplify $\frac{72}{108}$

The HCF is 36 so divide top and bottom by 36.

$72 \div 36 = 2$ and $108 \div 36 = 3$. So $\frac{72}{108} = \frac{2}{3}$

Example 2: $21 \div \frac{7}{8} = 21 \times \frac{8}{7} = \overset{3}{\cancel{21}} \times \frac{8}{\cancel{7}} = 24$

Exercise 20.2

Do these questions in your head, without using a calculator.
You may make jottings if necessary.

1 Write each of these fractions in its simplest form.

  a) $\frac{36}{48} =$ _____   b) $\frac{96}{108} =$ _____

CALCULATIONS AND MENTAL STRATEGIES 3

c) $\frac{270}{360} =$ _____

d) $\frac{175}{1000} =$ _____

e) $\frac{99}{451} =$ _____

f) $\frac{56}{77} =$ _____

g) $\frac{190}{570} =$ _____

h) $\frac{17}{68} =$ _____

i) $\frac{64}{80} =$ _____

j) $\frac{23}{69} =$ _____

k) $\frac{51}{68} =$ _____

l) $\frac{57}{95} =$ _____

2   Convert each of these decimals to a fraction in its simplest form.

a) 0.25 = _____

b) 0.025 = _____

c) 0.875 = _____

d) 0.0875 = _____

e) 0.6 = _____

f) 0.006 = _____

g) 0.05 = _____

h) 0.35 = _____

i) 0.37 = _____

j) 0.037 = _____

3   Work out these calculations. Cancel any common factors first and give your answers in their simplest form.

a) $\frac{1}{5}$ of 35 = _____

b) $\frac{2}{5}$ of 35 = _____

c) $\frac{2}{5}$ of 0.35 = _____

d) $\frac{1}{10}$ of 290 = _____

e) $\frac{3}{5}$ of 290 = _____

f) $\frac{7}{20}$ of 290 = _____

g) $\frac{1}{7}$ of 91 = _____

h) $\frac{6}{7}$ of 91 = _____

i) $\frac{3}{7}$ of 0.91 = _____

● CHAPTER 20

4 Work out these calculations. Give your answers in their simplest form.

a) $\frac{3}{11} - \frac{9}{11} + \frac{7}{11}$ _____

b) $\frac{6}{13} - 1 + \frac{9}{13}$ _____

c) $\frac{3}{4} + \frac{5}{8} - \frac{1}{2}$ _____

d) $1 - \frac{7}{9}$ _____

e) $2\frac{1}{2} + 1\frac{1}{4} - \frac{3}{4}$ _____

5 Work out these calculations. Give your answers in their simplest form.

a) $7 \times \frac{4}{5} =$ _____

b) $8 \times \frac{6}{7} =$ _____

c) $15 \div \frac{3}{4} =$ _____

d) $32 \div \frac{8}{9} =$ _____

e) $9 \times \frac{2}{3} \div 6 =$ _____

6 Work out these calculations.

a) $3 \div \frac{3}{7} =$ _____

b) $15 \div \frac{5}{2} =$ _____

c) $8 \div \frac{16}{9} =$ _____

d) $10 \div \frac{5}{8} =$ _____

e) $3 \div \frac{3}{4} =$ _____

f) $36 \div \frac{6}{7} =$ _____

g) $18 \div \frac{6}{7} =$ _____

h) $16 \div \frac{2}{5} =$ _____

i) $48 \div \frac{6}{11} =$ _____

j) $99 \div \frac{9}{10} =$ _____

k) $77 \div \frac{7}{8} =$ _____

l) $44 \div \frac{4}{9} =$ _____

Percentages

You need to be able to convert between fractions, decimals and percentages.

Exercise 20.3

1. Complete this table.

Fraction	Decimal	Percentage
$\frac{1}{2}$		
	0.25	
		75%
		12.5%
	0.375	
$\frac{5}{8}$		
		87.5%
$\frac{1}{10}$		

Fraction	Decimal	Percentage
$\frac{1}{5}$		
		30%
$\frac{2}{5}$		
		60%
	0.7	
$\frac{4}{5}$		
		90%

2. Write each of these percentages as a mixed number in its simplest form.

   a) 123% = _____

   b) 220% = _____

   c) 330% = _____

   d) 250% = _____

3. Write each of these fractions as a percentage.

   a) $\frac{2}{100}$ = _____

   b) $\frac{99}{1000}$ = _____

   c) $\frac{7}{10}$ = _____

   d) $\frac{7}{50}$ = _____

CHAPTER 20

4 Convert each of these percentages to a decimal.

a) 169% = _____

b) 750% = _____

c) 23% = _____

d) 1275% = _____

5 Convert each of these decimals to a percentage.

a) 1.03 = _____

b) 0.04 = _____

c) 0.875 = _____

d) 3.05 = _____

6 Work out these percentages.

a) 60% of 60 = _____

b) 125% of 36 = _____

c) 10% of 35 = _____

d) 12.5% of 40 = _____

e) 37.5% of 160 = _____

f) 160% of 37.5 = _____

g) 150% of 36 = _____

h) 36% of 150 = _____

i) 24% of 200 = _____

j) 80% of 125 = _____

k) 8% of 125 = _____

l) 125% of 8 = _____

Teacher comments

SECTION 4

22 Ratio and proportion

Ratio

A **ratio** shows the relative sizes of two numbers, similar to a **fraction**. To compare different ratios, it often helps to write each ratio as a percentage.

Exercise 22.1

1 Three ice-hockey players take penalty shots at goal. Player A scores 35 times from 50 shots, player B scores 42 times from 63 shots, and player C scores 40 times from 70 shots.

   a) What is each player's success rate as a percentage?

   Player A _____ Player B _____ Player C _____

   b) Write the players in order, most successful first.

   _____ _____ _____

2 $3000 is divided between three children in the ratio $1:2:3$. How much does each child receive?

   _____ _____ _____

3 Pedro is 160 cm tall. His friend Ahmet is 145 cm tall. What is the ratio of their heights in its simplest form?

   ____ : ____

● CHAPTER 22

4 A bag contains 750 balls. The ratio of the colours red to white to blue is $7:8:10$.

   a) Calculate the probability of choosing:

   (i) a blue ball _____

   (ii) not a red ball _____

   (iii) a blue, white or red ball. _____

   b) How many balls of each colour are there?

   _____

5 a) What is the ratio of the number of days in February to the number of days in a non-leap year?

   _____ : _____

   b) What is the ratio of the number of days in a year to the number of days in the first six months added together?

   _____ : _____

   c) Explain why the answer to part **b)** is not exactly $1:2$.

   _____

6 Write either 'yes' or 'no' to indicate whether each of the following pairs of ratios are equivalent.

   a) $7:11$ and $42:66$ _____   **b)** $8:3$ and $6:16$ _____

   c) $5:9$ and $35:28$ _____   **d)** $14:15$ and $30:28$ _____

   e) $14:15$ and $420:600$ _____   **f)** $19:27$ and $76:108$ _____

g) 85 : 15 and 16 : 3 _____ **h)** 85 : 15 and 3 : 7 _____

i) 1 : 3 and 3 : 1 _____ **j)** 12 : 7 and 156 : 91 _____

Direct proportion

A rate is a proportion that involves two different quantities. If the rate is constant, the two quantities are in **direct proportion**.

Exercise 22.2

Use proportion to work out these problems.
In each question assume that the rate stays the same.

1. Jose walks 12 km in 4 hours.
   How long will it take him to walk 19.5 km?

2. A machine lays 20 m of road in 30 minutes.
   How long will the machine take to lay 1 km of road?

3. 24 apples cost $2.40. What do 3 dozen apples cost?

4. In January, the exchange rates between US dollars, euros and pounds sterling were:

   $1 = €0.70 $1 = £0.61

   a) How many euros was $5000 worth? _____

   b) How many dollars was €1 worth? _____

   c) How many pounds was $250 worth? _____

   d) How many dollars was £1 worth? _____

   e) How many euros was £1 worth? _____

5 What is 35% of 450? _____

6 What percentage of 72 is 63? _____

7 What percentage of 50 is 45? _____

8 What is 450% of 35? _____

9 What percentage of 48 is 72? _____

10 What percentage of 40 is 60? _____

11 A girl scored 59.5 marks out of 85 in a test.
   What is her mark as a percentage?

   _____

12 75% of the 32 students in a class got an A in a test.
   How many students did not get an A?

   _____

13 36 out of the 40 students in a class come to school by bus.
   What percentage do not come by bus?

   _____

14 $900 is 75% of a man's monthly income.
   What is his monthly income?

   _____

Teacher comments

23 Functions and graphs

Linear functions

A **line** is made up of an infinite number of points. The **coordinates** of every point on a straight line all have a common relationship. In other words, the x and y values follow a pattern. It is this pattern that gives the **equation of the line**.

The equation $y = x - 1$ is an example of an **explicit** function. It gives the value of y directly in terms of x. The same equation can be re-written as $y - x + 1 = 0$. This is an **implicit** function.

To draw a graph from an equation that is written implicitly, it is usually easier to rearrange it into explicit form first.

Exercise 23.1

Plot each straight line on the grid provided.
First re-write the equation in explicit form and work out the coordinates of at least three points on the line.

1 $y - x = 3$

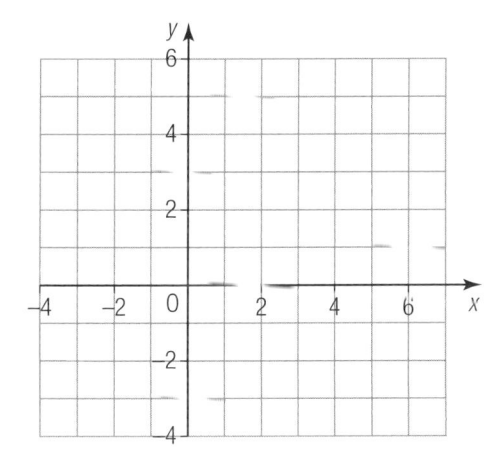

$y = $ _____

x					
y					

CHAPTER 23

2 $y - \frac{1}{2}x - 1 = 0$

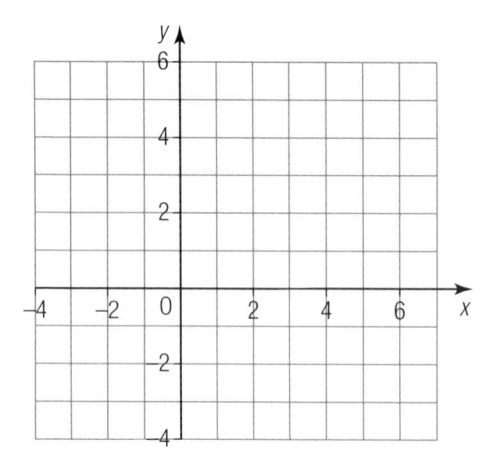

$y = $ _____

x					
y					

3 $2y = -x + 4$

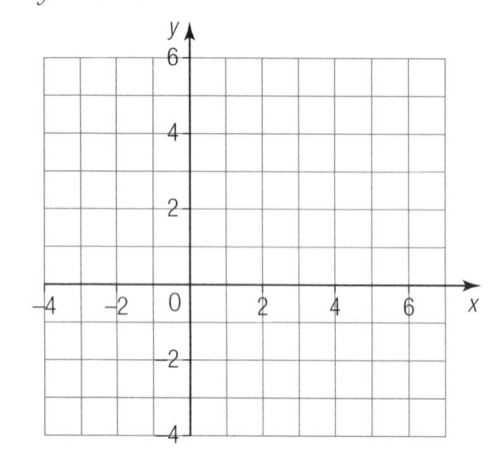

$y = $ _____

x					
y					

4 $-y + 2x - 2 = 0$

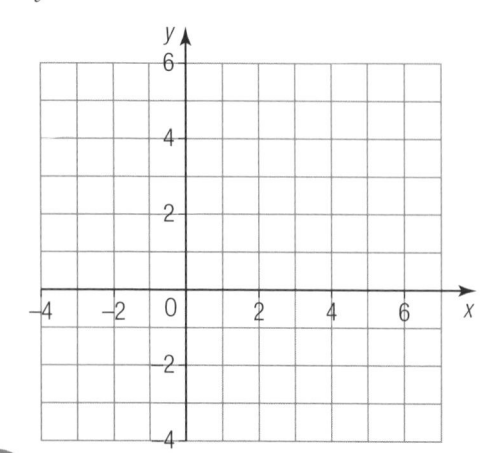

$y = $ _____

x					
y					

FUNCTIONS AND GRAPHS

The general equation of a straight line

The **gradient** of a line is a measure of how steep the line is. It is calculated by working out the difference between the *y* coordinates of two points on the line, divided by the difference between the *x* coordinates.

$$\text{Gradient} = \frac{y_2 - y_1}{x_2 - x_1}$$

Example 1: The gradient of the straight line between (6, 3) and (0, 6) is

$$\frac{y_2 - y_1}{x_2 - x_1} = \frac{6-3}{0-6} = \frac{3}{-6} = -\frac{1}{2}$$

Apart from straight lines that run parallel to the *y* axis, all straight lines cross the *y* axis at some point. This point is called the **y intercept**.

The equation of any straight line has the form $y = mx + c$, where *m* is the gradient and *c* is the *y* intercept.

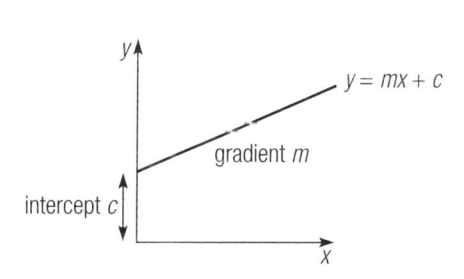

Example 2: The line with equation $y = 4x - 2$ has gradient 4 and *y* intercept −2.

If an equation is written implicitly, it is usually easier to rearrange it into explicit form before deducing its gradient and *y* intercept.

Exercise 23.2

1   Calculate the gradient of the straight line passing through each of these pairs of points.

   a) (2, 4) and (4, 8)   **b)** (2, 8) and (5, 9)

   Gradient = _____   Gradient = _____

   c) (5, 4) and (4, 7)   **d)** (−2, 13) and (2, 12)

   Gradient = _____   Gradient = _____

● CHAPTER 23

2 For each of the following graphs, work out the gradient, the y intercept and the equation of the straight line.

a)

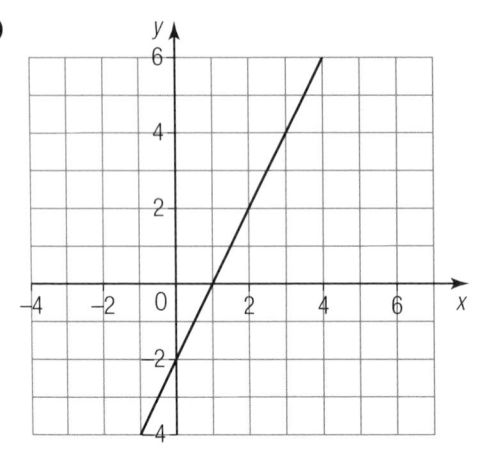

Gradient = _____

y intercept = _____

y = _____

b)

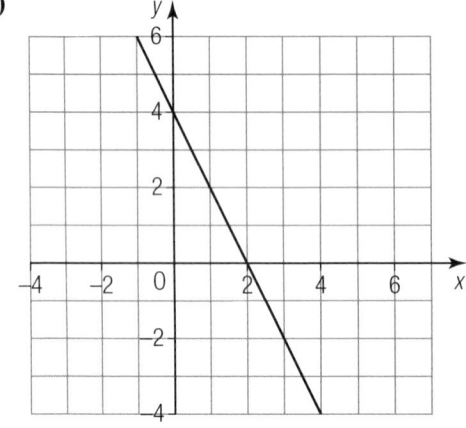

Gradient = _____

y intercept = _____

y = _____

c)

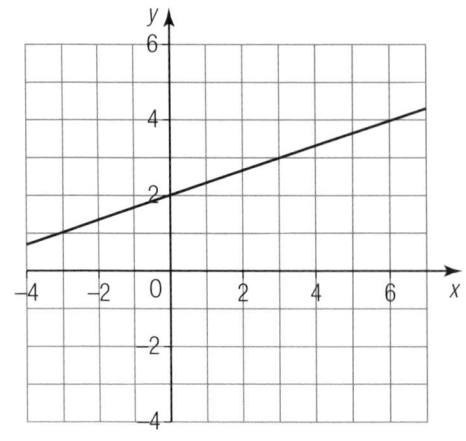

Gradient = _____

y intercept = _____

y = _____

FUNCTIONS AND GRAPHS

3 For the straight line represented by each of these equations, find the value of the gradient and the y intercept.

a) $y = 3x - 6$

b) $y = \frac{1}{3}x + 8$

Gradient = _____

Gradient = _____

y intercept = _____

y intercept = _____

c) $y = -6x + 8$

d) $y = 4$

Gradient = _____

Gradient = _____

y intercept = _____

y intercept = _____

e) $y - 4x = 0$

f) $y - x = -8$

Gradient = _____

Gradient = _____

y intercept = _____

y intercept = _____

g) $y + 3x = 7$

h) $y + \frac{1}{2}x = -4$

Gradient = _____

Gradient = _____

y intercept = _____

y intercept = _____

CHAPTER 23

i) $2y = 6x + 8$

j) $4y = -4x - 2$

Gradient = _____

Gradient = _____

y intercept = _____

y intercept = _____

k) $-3y = 12x + 3$

l) $5y = 10$

Gradient = _____

Gradient = _____

y intercept = _____

y intercept = _____

Inverse functions

The **inverse** of a function 'undoes' the effect of the original function. In function notation, the inverse of the function $f(x)$ is written as $f^{-1}(x)$.

Example: Find the inverse of the function $f(x) = 2x + 12$.

Write the equation in terms of y: $y = 2x + 12$
Swap x and y: $x = 2y + 12$
Rearrange to make y the subject: $2y = x - 12$
  $y = \frac{1}{2}x - 6$

The inverse function is $f^{-1}(x) = \frac{1}{2}x - 6$.

Exercise 23.3

1 Find the inverse of each of the following functions.

a) $f(x) = x - 8$

b) $f(x) = 3x - 12$

$f^{-1}(x) = $ _____

$f^{-1}(x) = $ _____

c) $f(x) = \frac{x-2}{4}$

d) $f(x) = \frac{2x-1}{6}$

$f^{-1}(x) =$ _____

$f^{-1}(x) =$ _____

e) $f(x) = \frac{1}{4}x - 1$

f) $f(x) = \frac{2}{3}x + 2$

$f^{-1}(x) =$ _____

$f^{-1}(x) =$ _____

Approximate solutions to equations

You can find an approximate solution to a pair of simultaneous linear equations by plotting the two lines on a graph. At the point where the two graphs intersect, the x and y values satisfy both equations and are the solution to the simultaneous equations.

Another way of finding an approximate solution to an equation is the method of **trial and improvement**. You choose an x value to substitute into the equation for the first trial and compare the result with what you need. You then adjust the x value for the next trial, getting closer and closer to the solution each time.

Example: One solution to the equation $x^2 + 3x = 3$ is between $x = 0$ and $x = 1$.

Use the method of trial and improvement to find this solution, giving your answer to one decimal place.

Try $x = 1$: $1^2 + 3(1) = 4$ 4 is too big, so try a smaller value of x.
Try $x = 0.5$: $(0.5)^2 + 3(0.5) = 1.75$ 1.75 is too small, so try a larger value.
Try $x = 0.8$: $(0.8)^2 + 3(0.8) = 3.04$ The value of x must be smaller than 0.8.
Try $x = 0.7$: $(0.7)^2 + 3(0.7) = 2.59$ The value of x must be bigger than 0.7.
The solution is between these two values.

We need a solution correct to one decimal place, so it is either $x = 0.7$ or $x = 0.8$.
Of these, $x = 0.8$ gives an answer closer to 3 than $x = 0.7$ does.
The solution to the equation is $x = 0.8$ (to one decimal place).

● CHAPTER 23

Exercise 23.4

1 For each of these pairs of equations, draw the two straight lines on the grid provided. Then use the coordinates of the point of intersection to find an approximate solution to the simultaneous equations.

   a) $y = \frac{1}{2}x - 3$ and $y = -3x + 2$

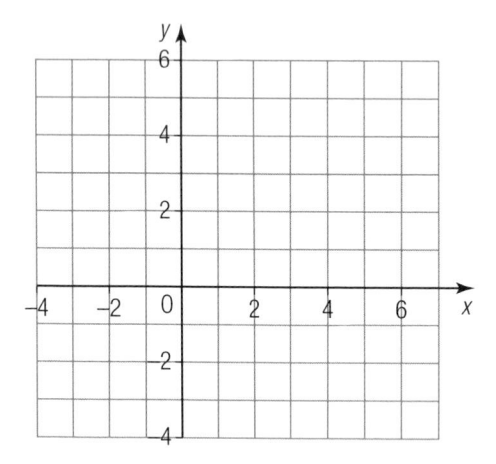

   $x =$ _____ and $y =$ _____

   b) $y = -\frac{1}{3}x + 3$ and $y = 2x + 4$

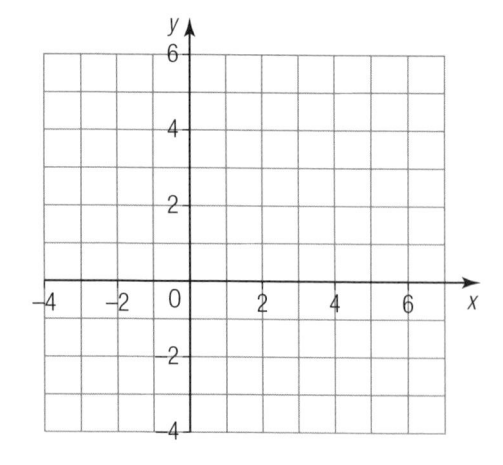

   $x =$ _____ and $y =$ _____

FUNCTIONS AND GRAPHS

2 One solution of the equation $2x^2 - 3x - 4 = 0$ is between $x = 0$ and $x = 5$. Use the method of trial and improvement to find this solution, giving your answer correct to one decimal place.

$x =$ _____

3 The equation $-x^2 - 3x = -3$ has two solutions between $x = -5$ and $x = 5$. Use the method of trial and improvement to find both solutions, giving your answers correct to the nearest integer.

$x =$ _____

● CHAPTER 23

Graphs from real-life situations

You can often solve a real-life problem by using the information given to form an equation and then drawing a graph.

Exercise 23.5

1 Cloth for making curtains is sold by length. The cost is $8.20 per metre.

   a) Write a formula for calculating the cost (C) of buying x metres of cloth.

   _____

   b) Plot a graph of x against C.

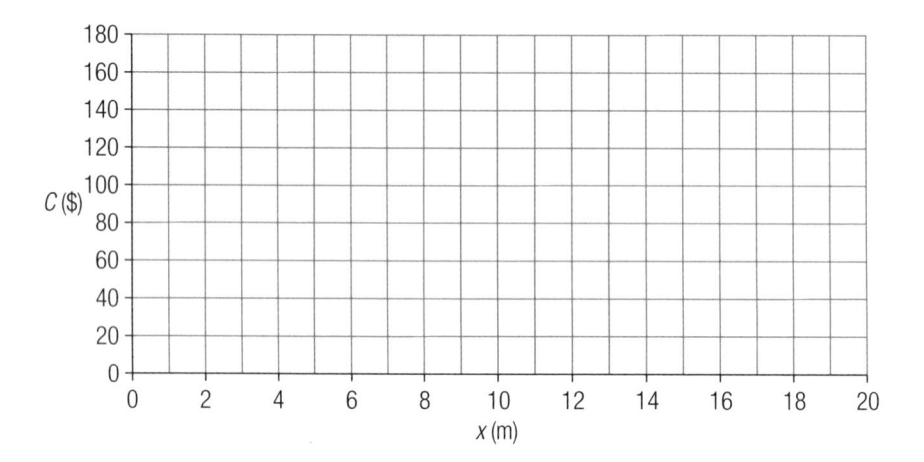

   c) Use your graph to estimate how much cloth was bought, if the total cost was $120.

   _____

FUNCTIONS AND GRAPHS

2 Two electricity companies have different tariffs:

Company X charges $30 per month and then $0.02 for each unit of electricity used.

Company Y charges $12 per month and then $0.04 for each unit of electricity used.

a) For each company, write a formula that shows the total monthly bill (C) if n units of electricity are used.

Company A _____

Company B _____

b) Plot a graph of C against n for both companies on the same axes.

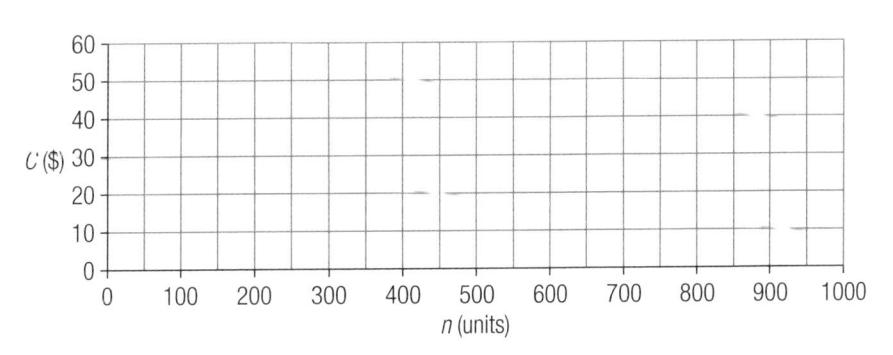

c) Use your graph to estimate the number of units for which both companies would charge the same amount.

_____

Teacher comments

24 Bearings and drawings

Maps and scale diagrams

The **scale** of a map is often given as a ratio. For example, a scale of 1:100 000 means that 1 cm on the map represents 100 000 cm on the ground.
Plans are also drawn to a scale, which is usually expressed in the same way.

Exercise 24.1

1 A scale plan of a room is to a scale of 1:60.
   The room is 12 cm long on the plan.
   How long is the real room?

   _____

2 A model boat is 25 cm long. The scale is 1:150.
   How long is the real boat?

   _____

In questions 3 and 4, give your answers in a suitable unit.

3 A map has a scale of 1:50 000.
   How far is a journey which measures 35 cm on the map?

   _____

4 A map has a scale of 1:25 000.
   How many centimetres on the map is a journey of 5.5 km?

   _____

BEARINGS AND DRAWINGS

Bearings

In a **three-figure bearing**, a direction is given in degrees measured clockwise from north, which has the bearing zero.

To measure the bearing from X to Y, draw a north arrow at X, measure the angle between the north arrow and the line XY in a clockwise direction and write it as a three-digit number.

If you know the bearing from X to Y, you can calculate the bearing from Y to X. This is called the **back bearing**. The difference between a bearing and its back bearing is always 180°.

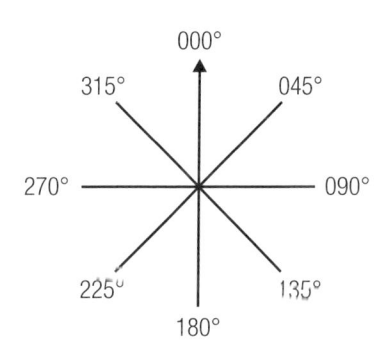

Exercise 24.2

In questions 1–3, use a protractor to measure the bearings. Show the north arrow in each case.

1

• A   • B

Bearing from A to B = _____

2

• Y

    • X

Bearing from X to Y = _____

119

● CHAPTER 24

3

•C

•A

•B

Bearing from A to B = _____

Bearing from C to B = _____

Bearing from A to C = _____

4 Calculate the back bearing for each of these bearings.

	Bearing from X to Y	Back bearing from Y to X
a)	055°	
b)	108°	
c)	355°	
d)	180°	

5 a) A car starts at a point *A*. It travels a distance of 5 km on a bearing of 135° to point *B*. From *B* it travels 8 km on a bearing of 280° to point *C*.
Draw a diagram to show these bearings and journeys. Use a scale of 1 cm : 1 km and take north as a line vertically up the page.

•
A

b) The car makes its way straight back from *C* to *A*.

   (i) What distance does it travel? _____

   (ii) On what bearing does it travel? _____

c) Another car travels directly from *A* to *C*.

   (i) What distance does it travel? _____

   (ii) On what bearing does it travel? _____

● CHAPTER 24

6 The map extract shows a part of Malaysia and Singapore.
   The scale of the map is 1 : 4 000 000.

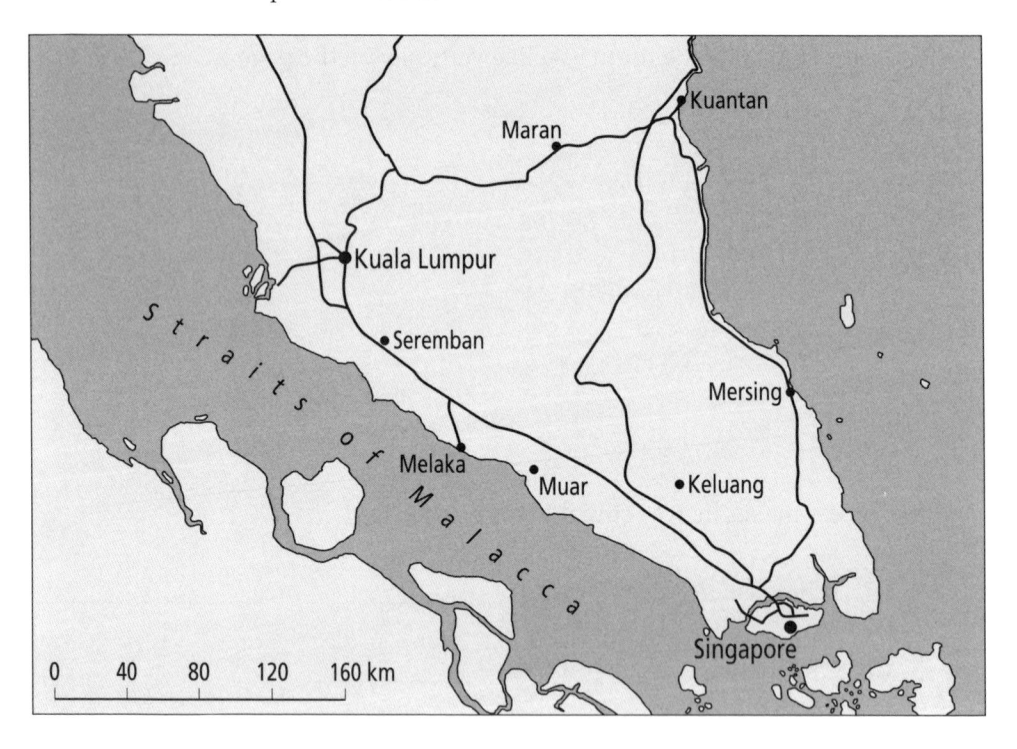

   a) A tourist travels from Singapore to Kuala Lumpur and then on to Kuantan. By measuring the map with a ruler, calculate the real (direct) distance in km that the tourist travels in total.

   _____

   b) What is the bearing from

      (i) Singapore to Kuala Lumpur? _____

      (ii) Kuala Lumpur to Kuantan? _____

Simple loci

A **locus** (plural **loci**) describes where a series of points (or a single point) lie, where the points fit a particular rule.
Most simple problems involving loci involve one of these types:

- points at a fixed distance from a given point – this is the circle with that radius and centred on the point

BEARINGS AND DRAWINGS

- points at a fixed distance from a straight line

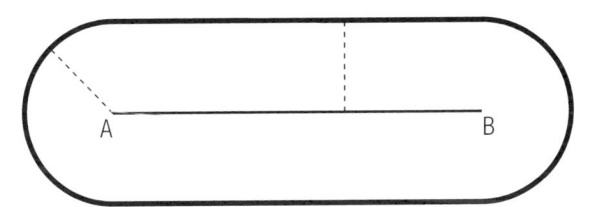

- points at the same distance from two points – this is the perpendicular bisector of the line between the two points.

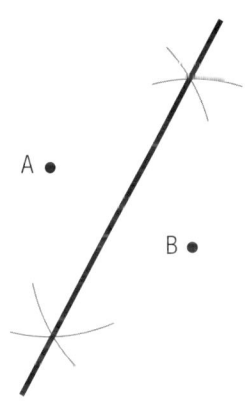

These simple loci are straight or curved lines, but a locus can also be a region or area.

Exercise 24.3

1 Construct the locus of all the points that are 2 cm from the point X.

X •

● CHAPTER 24

2 Construct the locus of all the points that are the same distance from points X and Y.

X •

• Y

3 This diagram shows a plan view of a rectangular garden. The scale is 1 cm : 1 m. The wall of the house runs along one side of the garden.

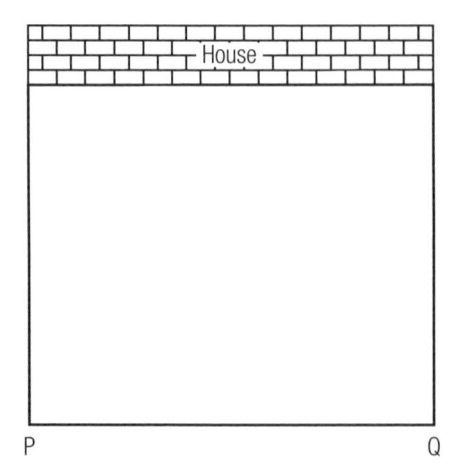

House

P   Q

The owner of the house wants to grow grass in the garden. The grass must be *more* than 1.5 m from the side of the house and *at least* 2 m from the corners P and Q.

On the diagram, shade the locus of all the points where the grass can be grown.

4 This diagram is a plan view of a girl and a boy standing on the same side of a brick wall. The wall is taller than they are, so they cannot see over it.

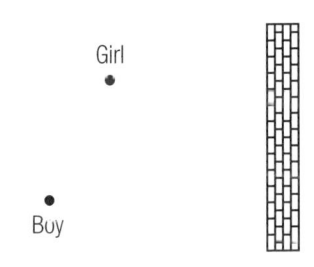

Girl

Boy

a) On the diagram, show the locus of all the points which the boy can see but the girl cannot see. Label this region *B*.

b) Show the locus of all the points which the girl can see but the boy cannot see. Label this region *G*.

Teacher comments

25 Measures and the circle

The circle

The circumference of any circle is given by the formula:

Circumference = π × 2 × radius   or $C = 2\pi r$

The area of a circle can be calculated using the formula:

Area of a circle = πr^2

Exercise 25.1

1   Calculate the circumference of each of these circles.
    Give each answer as a multiple of π.

a) 2 cm

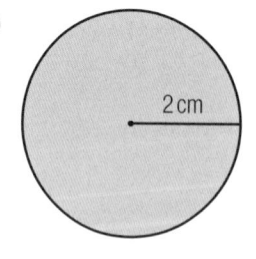

b) 4 cm

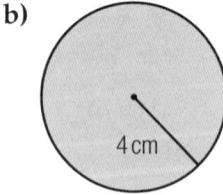

Circumference = _____

Circumference = _____

MEASURES AND THE CIRCLE

c) 37 mm

d) 7.4 m

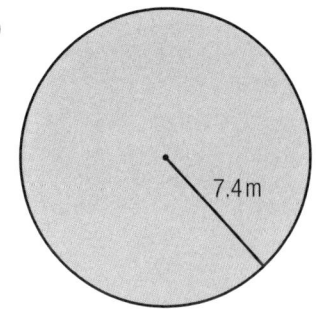

Circumference = _____   Circumference = _____

2 Calculate the area of each of these circles. Give each answer as a multiple of π.

a) 3 cm

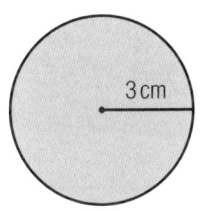

b) 0.3 cm

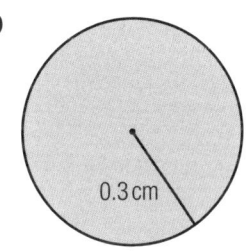

Area = _____   Area = _____

c) 2.7 mm

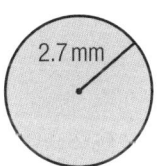

d) 27 cm

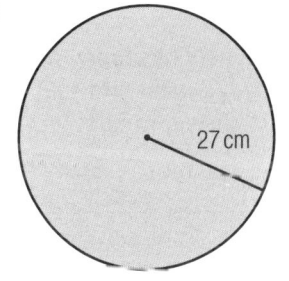

Area = _____   Area = _____

● CHAPTER 25

e)

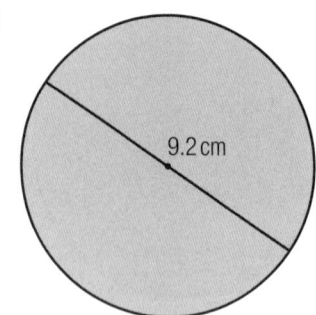

9.2 cm

f)

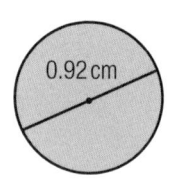

0.92 cm

Area = _____   Area = _____

3 A car wheel has an outer radius of 22 cm.

    a) Calculate how far the car has travelled after one complete turn of the wheel. Give your answer correct to two decimal places.

    _____

    _____

    b) How many complete turns does the wheel make in a journey of 2π km? Give your answer correct to the nearest whole number.

    _____

    _____

4 A bicycle wheel has diameter 64 cm.
How far will a cyclist travel if the wheel rotates 1000π times?
Give your answer in kilometres and correct to one decimal place.

_____

_____

MEASURES AND THE CIRCLE

5 A circular hole of radius 20 cm is cut from a disc of radius 22 cm to produce a ring as shown.
   Calculate the cross-sectional area of the ring.
   Give your answer correct to one decimal place.

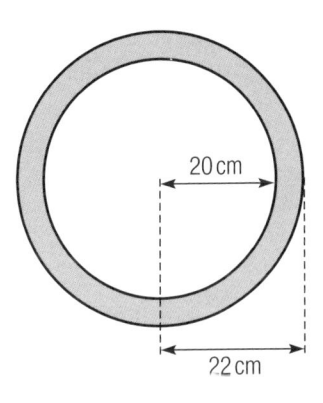

   _____

   _____

6 Four circles of radius 3 cm are drawn side by side, just touching. They just fit inside a rectangle.

   a) Draw a sketch to show the circles and rectangle.

   b) What is the area of the rectangle?

   _____

   c) What is the area of each circle? Give your answer as a multiple of π.

   _____

   d) Calculate the area inside the rectangle which is not covered by the circles.
      Give your answer correct to one decimal place.

   _____

   _____

● CHAPTER 25

7 A garden is made up of a 20 m by 10 m rectangle which is planted with grass and two semicircular vegetable patches, one on each short end of the rectangle.

   a) Draw a sketch of the garden.

   b) Calculate the perimeter of the garden.
      Give your answer correct to one decimal place.

   _____

   c) Calculate the total area of the garden.
      Give your answer correct to one decimal place.

   _____

Prisms

The volume of any prism is given by the formula:

   Volume of a prism = area of cross-section × length

In a cylinder, the cross-section is a circle, so the volume is:

   Volume of a cylinder = $\pi \times r^2 \times$ length

The surface area of any three-dimensional shape is the total area of all of its faces. In a cylinder, the surface area is made of two circles and a rectangle. The length of the rectangle is the same as the circumference of one of the circles.

Exercise 25.2

1 An apartment block is a cuboid.
   It measures 120 m by 50 m and is 30 m high.

   a) Mark these dimensions on the diagram.

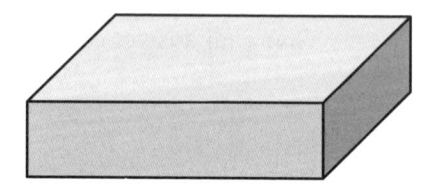

MEASURES AND THE CIRCLE

b) (i) Calculate the volume of the apartment block.

_____

(ii) Calculate the area of ground it covers.

_____

(iii) Calculate the total area of the outside walls and the roof.

_____

_____

c) 80% of the volume of the block is apartments. The rest is made up of stairs and other shared areas. Calculate the total volume of the apartments.

_____

2 At a refinery, petrol is stored in a large tank in the shape of a cylinder. The exposed surface of the cylinder is made of a circle and a rectangle. The length of the rectangle is the same as the circumference of the circle.

50 m

90 m

a) Mark the three missing dimensions on this diagram of the net.

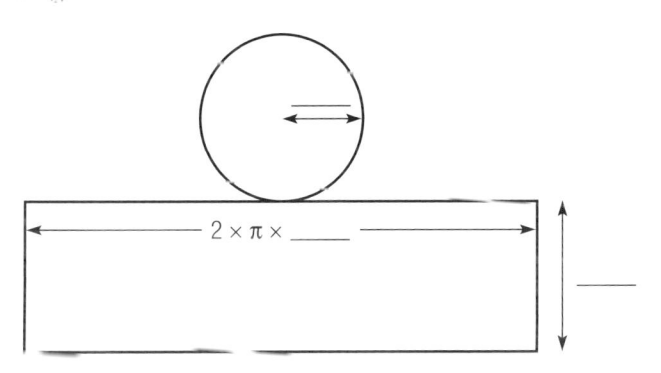

2 × π × _____

b) Calculate the volume of the tank. Give your answer as a multiple of π.

_____

131

c) Calculate the visible surface area of the tank.
Give your answer as a multiple of π.

_____

d) The petrol is transported by trucks.
Each truck has a cylindrical tank of length 15 m and radius 2.5 m.
Calculate the volume of petrol in each of these smaller tanks.
Give your answer as a multiple of π.

_____

e) How many trucks would be needed to empty a full large petrol tank?

_____

_____

3 The volume of this cuboid is 6750 cm³.

  a) Write an equation for the volume of the cuboid in terms of x.

  _____

  b) Solve the equation to find the value of x.

  $x =$ _____

  c) Calculate the surface area of the cuboid.

  Surface area = _____

MEASURES AND THE CIRCLE

4 The total surface area of a cylinder is given by the formula

$A = 2\pi r(r + l)$

where r is the radius of the cross-section and l is the length of the cylinder.
This cylinder has a radius of 6 cm and a total surface area of 192π cm².

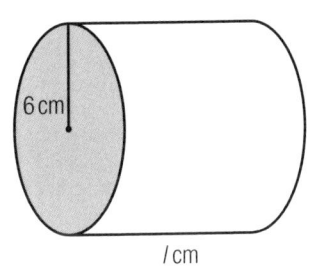

6 cm

l cm

a) Write an equation for the total surface area of the cylinder in terms of l.

_____

b) Solve the equation to find the value of l.

$l =$ _____

c) Calculate the volume of the prism.
Give your answer correct to one decimal place.

Volume = _____

Teacher comments

133

26 Probability

Successive events

Successive events are events that happen one after the other. One way of displaying all the possible outcomes is to use a **sample space diagram**.

Exercise 26.1

1. Two dice are rolled. Dice A has four sides and is numbered 1–4; dice B has eight sides and is numbered 1–8. The scores of the two dice are multiplied together.

   a) Complete the sample space diagram to show all the possible outcomes.

		\multicolumn{8}{c}{Dice B}
		1	2	3	4	5	6	7	8
Dice A	1								
	2			6					
	3						18		
	4								

   b) How many possible outcomes are there? _____

   c) What is the probability of getting a score of 8? _____

   d) What is the probability of not getting a score of 8? _____

   e) Which score has the greatest probability of occurring? _____

PROBABILITY

2 Two spinners are spun. Spinner A has eight sections numbered 1–8; spinner B has four sections, coloured red, yellow, blue and white.

a) Draw a sample space diagram to show all the possible outcomes.

b) How many possible outcomes are there? _____

c) What is the probability of getting a 4 and yellow? _____

d) What is the probability of not getting a 4 and yellow? _____

e) What is the probability of getting an even number and red? _____

f) What is the probability of getting an even number *or* red? _____

Relative frequency and probability

In real life, the experimental probability is often not the same as the theoretical probability.
Another name for the experimental probability is **relative frequency**.

$$\text{Relative frequency} = \frac{\text{number of successful trials}}{\text{total number of trials}}$$

Exercise 26.2

1. A four-sided dice, numbered 1–4, is suspected of being biased.

   a) If a fair four-sided dice is rolled 20 times, approximately how many times would you expect each number to come up?

   _____

   b) The dice is rolled 20 times. The table shows the results.

Number	1	2	3	4
Frequency	5	5	4	6

   From these results, is the dice likely to be biased? Justify your answer.

   _____ because _____

   _____

   c) The dice is rolled a further 80 times. The table shows all 100 results.

Number	1	2	3	4
Frequency	15	60	10	15

   From these results, is the dice likely to be biased? Justify your answer.

   _____ because _____

   _____

   d) Which of the two sets of results is more reliable? Justify your answer.

   _____ because _____

   _____

   e) Approximately how many times would you expect the number 4 to come up if it is rolled 1000 times? (Assume that the larger set of results gives a true representation of the dice.)

   _____

f) The dice is rolled x times and the number 2 comes up 900 times. Calculate the value of x.

2 A pharmaceutical company is conducting a trial on a new drug. 400 patients with the same condition are given the new drug and 400 patients are given a placebo. The tables show the results of the trial.

New drug

Patient condition after treatment	Got better	Stayed the same	Got worse
Frequency	280	40	80

Placebo

Patient condition after treatment	Got better	Stayed the same	Got worse
Frequency	120	250	30

a) What is the experimental probability of a patient getting worse if they are given the new drug?

b) What is the experimental probability of a patient getting better if they are given the placebo?

c) The pharmaceutical company claims that the results show that the new drug works. Comment on this statement in the light of the results.

Teacher comments

27 Calculations and mental strategies 4

The order of operations

The order of priority in calculations is:

Brackets
Indices
Division and/or **M**ultiplication
Addition and/or **S**ubtraction

You can use the shorthand **BIDMAS** to help you remember this.

Exercise 27.1

1 Write in any brackets which are needed to make each of these calculations correct.

   a) $12 - 8 \times 3 = 12$

   b) $5 \times 5 + 4 = 45$

   c) $12 \times 3 + 9 - 4 = 96$

   d) $10 - 14 \times 13 + 3 = -64$

   e) $9 + 6 - 3 \div 3 + 4 = 18$

   f) $19 + 6 - 66 \div 2 + 10 = 2$

   g) $20 - 48 \div 2 + 6 = 14$

   h) $20 - 48 \div 2 + 6 = 2$

   i) $40 - 8 \div 2 + 6 = 22$

   j) $20 - 12 \div 2 + 6 = 10$

   k) $20 - 8 \div 2 + 6 = 22$

   l) $8 + 3 \times 4 - 6 = 14$

   m) $8 + 3 \times 4 - 6 = 38$

   n) $8 + 3 \times 4 - 6 = -22$

   o) $8 + 3 \times 4 - 6 = 2$

2 Work out these calculations without using a calculator.
   You may make jottings if necessary.

   a) $\dfrac{(4+20)}{2^3} = $ _____

   b) $\dfrac{(30+24)}{3^3} = $ _____

c) $\dfrac{(7+18)}{5^2} =$ _____

d) $\dfrac{(32+32)}{2^4} =$ _____

e) $2^2 + (5^2 + 1) =$ _____

f) $(4^2 + 4) \times 2 - 2^2 =$ _____

g) $10 + (-2) - 2^4 =$ _____

h) $2^5 - 3^3 =$ _____

i) $2 + \dfrac{(25-9)}{4} + (10 - 3^2) =$ _____

j) $\dfrac{6^3}{(16+2)} + (25 + 5^2) =$ _____

Multiplying and dividing by numbers between 0 and 1

You can use multiplication facts you know to estimate the answers to more complicated multiplications and divisions. Remember that:

- multiplying by a number greater than 1 increases the value
- multiplying by a number less than 1 decreases the value
- dividing by a number greater than 1 decreases the value
- dividing by a number less than 1 increases the value.

Exercise 27.2

Estimate the answer to each of these calculations.
In each question, use your estimate from part **a)** to estimate the answer to part **b)**.

1 a) 8.24×5 _____

   b) 8.24×0.05 _____

2 a) $8.24 \div 5$ _____

   b) $8.24 \div 0.05$ _____

● CHAPTER 27

3 a) 2.9×2 _____

   b) 2.9×0.02 _____

4 a) $2.9 \div 2$ _____

   b) $2.9 \div 0.02$ _____

5 a) 0.32×4 _____

   b) 0.32×0.4 _____

6 a) $0.32 \div 4$ _____

   b) $0.32 \div 0.4$ _____

7 a) $6 \div 8$ _____

   b) $6 \div 0.08$ _____

8 a) 6×8 _____

   b) 600×0.008 _____

9 a) 60×2 _____

   b) 600×0.02 _____

10 a) $30 \div 2$ _____

   b) $300 \div 0.2$ _____

Teacher comments